D0773161

Diamonds

Resources Series

Gavin Bridge & Philippe Le Billon, *Oil*

Jennifer Clapp, *Food*

Peter Dauvergne & Jane Lister, *Timber*

Elizabeth R. DeSombre & J. Samuel Barkin, *Fish*

David Lewis Feldman, *Water*

Derek Hall, *Land*

Michael Nest, *Coltan*

Ian Smillie

polity

First published in 2014 by Polity Press

Polity Press
65 Bridge Street
Cambridge CB2 1UR, UK

Polity Press
350 Main Street
Malden, MA 02148, USA

ISBN-13: 978-0-7456-7230-4
ISBN-13: 978-0-7456-7231-1(pb)

A catalogue record for this book is available from the British Library.

Typeset in 10.5 on 13pt Scala by
Servis Filmsetting Ltd, Stockport, Cheshire
Printed and bound in Great Britain by Clays Ltd, St Ives plc

Contents

Abbreviations

AFDL	Alliance of Democratic Forces for the Liberation of Congo-Zaire (Alliance des forces démocratiques pour la libération du Congo)
CAR	Central African Republic
CSR	corporate social responsibility
DDI	Diamond Development Initiative
DNPM	National Department of Mineral Production (Departamento Nacional de Produção Mineral)
DRC	Democratic Republic of Congo
DRI	Directorate of Revenue Intelligence
DTC	Diamond Trading Company
FNLA	National Liberation Front of Angola (Frente Nacional de Libertação de Angola)
GATT	General Agreement on Tariffs and Trade
HRD	Diamond High Council (Hoge Raad voor Diamant)
IBA	impact benefit agreement
IDSO	International Diamond Security Organisation
JRC	Responsible Jewellery Council
KP	Kimberley Process
KPCS	Kimberley Process Certification Scheme
LAICPMS	laser ablation inductively coupled plasma mass spectrometry
LVMH	Louis Vuitton Moët Hennessy
MIBA	Société Minière de Bakwanga

MLC	Congolese Liberation Movement (Mouvement de libération du Congo)
MONUA	UN Observer Mission in Angola
MONUSCO	United Nations Stabilization Mission in the Democratic Republic of Congo (Mission de l'Organisation des Nations Unies pour la stabilisation en République démocratique du Congo)
MPLA	Popular Movement for the Liberation of Angola (Movimento Popular de Libertação de Angola)
NDMC	National Diamond Mining Company
NGO	non-governmental organization
NPLF	National Patriotic Liberation Front
OFAC	Office of Foreign Assets Control
OSLEG	Operation Sovereign Legitimacy
PAC	Partnership Africa Canada
ROC	Republic of Congo
RUF	Revolutionary United Front
SLST	Sierra Leone Selection Trust
UAE	United Arab Emirates
UNAVEM	United Nations Angola Verification Mission
UNGA	United Nations General Assembly
UNITA	Union for the Total Independence of Angola (União para la Indepêndencia Total de Angola
UNMIL	United Nations Mission in Liberia
UNOCI	United Nations Operation in Côte d'Ivoire (Opération des Nations Unies en Côte d'Ivoire
UNSC	United Nations Security Council
WDC	World Diamond Council
WTO	World Trade Organization
ZANU-PF	Zimbabwe African National Union-Patriotic Front

Introduction

I first encountered diamonds in 1967 when, fresh out of university, I went to Sierra Leone to teach at a small secondary school in the heart of the country's remote diamond district. Koidu was a wild and lawless place – not, perhaps, unlike Dawson City at the height of the Yukon gold rush. But without snow.

After my time in Koidu, I didn't give diamonds much thought, going on to work elsewhere in Africa and Asia as a development practitioner and later as an aid administrator, consultant and sometime writer. Then, during the late 1990s, as Sierra Leone descended into one of the world's most horrific humanitarian crises, diamonds came into focus again when I joined an effort to understand how the war was being financed. My colleagues and I learned that diamonds were also fueling conflict in other countries. In 2000, as Sierra Leone's war entered its ninth year, I was appointed to a UN Security Council Expert Panel to examine the connection between diamonds and weapons. I traveled extensively to the diamond capitals of the world: Antwerp, London, New York, Tel Aviv; and to Freetown, Monrovia, Conakry, and Johannesburg – places where diamonds began their journey through a secretive underground network that ran from rebel armies to the fingers of brides in waiting.

I took part in the "blood diamond" campaign and I participated in negotiations that led to the creation in 2003 of the first-ever international certification system for rough diamonds. In 2008 I was the first witness at the war crimes trial in The Hague of former Liberian President and warlord, Charles Taylor, where I spoke about his role in the illicit diamond trade. I helped to start an organization called the Diamond Development Initiative that works on the problems of Africa's many artisanal diamond diggers. And in 2010 I wrote a book about diamonds called *Blood on the Stone: Greed, Corruption and War in the Global Diamond Trade.*

I will explain below why the content of this book is different from that one, but first a note on style. Rather than write this book entirely in the third person, I have been encouraged by Polity Press to describe in the first person some of the events in which I played a direct role. I hope the approach will make the book readable and a little more genuine than if I were to feign distance and complete impartiality.

There are several compelling reasons for a new book about diamonds, one that brings together for the first time three aspects of the diamond industry: the diamond mystique, born of geology, history, and commercialization; blood diamonds; and the development potential in very poor countries of a mineral sold on the basis of love, prestige, and wealth.

The diamond "industry" involves some of the world's largest mining companies (De Beers, Rio Tinto, Anglo American); millions of artisanal diggers, cutters, and polishers; and a $70 billion retail jewelry business. Polished diamonds are India's largest export, rough diamonds are the largest export from the Democratic Republic of Congo, and diamonds represent significant elements in the economies of Australia, Belgium, Botswana, Canada, Israel, Russia, and South Africa. The irony in the numbers is that the sole use of gem diamonds is for decoration. They have no other purpose. Their mystique has deep historical roots, but it is today a product more of Hollywood, advertising, and market management than it is of reality, or even scarcity.

Because of their great value, diamonds have always been of interest to thieves, smugglers, and the entertainment industry. This aspect alone makes them worthy of study, and over the years several popular books have been written about diamonds: Edward Jay Epstein's 1982 *The Rise and Fall of Diamonds: The Shattering of a Brilliant Illusion*; *The Last Empire: De Beers, Diamonds and the World,* by Stefan Kanfer (1993); *Diamond: A Journey to the Heart of an Obsession* by Matthew Hart (2001). Diamonds have been a plot device and the *leitmotif* for novelists from Rider Haggard to Graham Greene and Ian Fleming, and songwriters from Jule Styne to Kanye West.

There is much more than this to the diamond story, however. During the 1990s, rebel armies in Angola, Sierra Leone, and the Democratic Republic of Congo – bereft of Cold War financial support – turned to the exploitation of natural resources to finance their wars. Diamonds soon became the most expedient vehicle for purchasing illicit weapons in a post-Cold War world awash in cheap AK-47s. "Blood diamonds" fueled wars that took the lives of hundreds of thousands of Africans, eventually becoming the focus of attention for humanitarian organizations, campaigning non-governmental organizations (NGOs), governments, and the UN Security Council. This aspect of the industry also became a subject of popular novels and films, including the Hollywood thrillers *Die Another Day* (2002), *Lord of War* (2005), and *Blood Diamond* (2006). Serious books were written on the subject as well, among them *Blood Diamonds* (Greg Campbell, 2002, updated in 2012) *Blood from Stones* (Douglas Farah, 2004), and Tom Zoellner's *The Heartless Stone* (2006).

The campaign to curtail the trade in conflict diamonds is of interest for several reasons. First, it involves the hard work by dedicated NGOs, journalists, politicians, business leaders, and civil servants to create a regulatory system in an industry that had successfully defied US anti-trust legislation and government controls right through a century that included two world wars, the Cold War, and the end of colonialism. The second theme has to do with regulation. Completely unregulated before 2003, diamonds had become the perfect tool for money laundering, tax evasion, drug-running, and weapons-trafficking. The regulatory system that was eventually initiated in 2003 is of interest because of the negotiations that brought industry, civil society, and 81 governments together at the same table.[1] It is important because the Kimberley Process Certification Scheme was viewed as a possible model for the regulation of other minerals that have fallen prey to rebel armies in Africa – gold, tungsten, tantalum, and tin. Third, the Kimberley Process is of continuing interest because, despite its initial promise, it has foundered on political and commercial shoals that have, astonishingly, and despite the best efforts of some participating governments and NGOs, seen it condone corruption, human rights abuse, smuggling, and violence. This aspect of the diamond story is very much alive and current, but it is rarely examined outside the gray literature produced by interested NGOs and industry. The implications of a failed Kimberley Process are, for African producing countries and the industry, enormous.

These represent three aspects of the diamond story, but there is a fourth that has rarely been examined: the role and potential of diamonds as a generator of development in some of the world's poorest countries. Geology has scattered diamonds in a very democratic manner. By volume, Australia has in recent decades been the world's largest producer. Some of the most valuable mines are found in Russia and Canada. India, the sole source of diamonds in ancient times, is today home to – by some estimates – almost a million diamond cutters and polishers. Brazil, Venezuela, and Guyana host small but not unimportant diamond operations. But more than half of the world's diamonds, by value, are produced in

Africa. Botswana is by far the largest producer, and diamonds have played a major role in South Africa and Namibia. They represent key elements in the economies of 13 other African countries, among them the poorest in the world. But diamonds, which seemed to hold great promise during the first half of the twentieth century, were whisked away under the noses of colonial governments willing to settle for a pittance of their value in royalties. The independence movement and its outcome in many countries led diamonds into darker paths of corruption, theft, money laundering, and eventually violence.

Today, things have changed. Despite its failings, the Kimberley Process has turned a bright light on diamonds. There is much greater transparency. And an industry once insulated by a cartel and its compliant customers has been obliged to do business differently. The same is true of certain African elites who once saw diamonds as their private milch cow. In many developing countries, diamonds were at the best of times, economically speaking, a zero-sum game, and at the worst of times the center of cataclysmic violence. Today, they represent a fragile but renewed opportunity for development. This part of the diamond story has rarely been told.

I am very grateful for the helpful comments of people who read an early draft: Dorothée Gizenga who has a development perspective on the diamond industry, Alan Martin who has advocacy experience, and Matt Runci, a long-time industry insider with an ethical perspective on the long haul. Shawn Blore, diamond investigator *par excellence*, dug up much of the detail on which this book is based and persuaded me to toughen up the parts of chapter 6 where I might have been going soft.

Ian Smillie
Ottawa

CHAPTER ONE

The Geology and History of Diamonds

Geology

The geology of diamonds is important to an understanding of how the industry has developed, why blood diamonds became so ubiquitous, why some mines require hundreds of millions of dollars in capital investment, and why in other places diamonds can be mined with little more than a shovel, a sieve, and a strong back.

Most of the diamonds mined today were formed more than 100 million years ago, some perhaps 3 billion years ago, long before the dinosaurs, long before single-cell organisms began to turn themselves into what we would recognize today as animal life. Diamonds formed in the upper mantle of the earth, some 150 kilometers below the surface, between its core and its crust. The right combination of minerals, heat, and pressure formed the crystals, and volcanic eruptions through the crust brought them to the surface, embedded in what has become known as kimberlite magma. Some of these "pipes" exploded into the atmosphere, scattering magma and diamonds across vast areas. Others never made it to the surface, leaking into horizontal "dykes," sometimes a kilometer or more in length. Yet others saw daylight and then sank back, soon enough covered – in the millennial sense of the term "soon enough" – in silt and debris, hidden from view.

Kimberlite eruptions took place around the world, egalitarian in their geographic spread. The best in terms of their diamond content are in Southern Africa, northern Canada and Yakutia in Russia's far northeast. But diamonds are also found in Australia and West Africa, and directly across the Atlantic from West Africa in Brazil, where the South American land mass broke away from Africa 200 million years ago.

Kimberlite pipes are petrified conical formations with relatively small footprints. The smallest have a surface area of only 5 or 6 acres, while the largest, like one at Fort à la Corne in Saskatchewan, cover as many as 500 acres. Not all of these pipes were created equal. The quality of their diamonds, if there are any at all, depends on a variety of factors, including the speed with which the volcanic magma rose to the surface. Thousands of kimberlite pipes have been discovered, but the number worthy of investment – the ones that are economically feasible – can be counted on the fingers of four or five hands. Given the level of effort and expenditure that goes into exploration, however, it is clear that geologists believe there are more to be found. Canada is a reminder and an icon. The pipes that led to the creation of the great diamond mines in Canada's Northwest Territories lay undiscovered until 1991, even though the search for them – a modern-day quest for King Solomon's Mines – had consumed tens of millions of dollars over the previous two decades.

Time has changed what nature wrought. Over several ice ages and 100 million rainy seasons, the tops of some kimberlite pipes eroded, and the diamonds or their trace elements were transported hundreds of miles from their origin. Telltale indicator garnets may be close to the original pipe, or they may have been carried hundreds of miles away by rivers, or in a jumble of glacial moraine. In the case of some well-established diamond areas, the originating kimberlite pipe still remains to be found. The world's first, and some of its best, diamonds were mined at Golconda in central India, and yet their geological source has never been discovered.

Erosion has created what are known as alluvial diamonds, from the Latin *alluvium*, meaning "to wash against." Alluvial diamonds have washed away from the original kimberlite pipe, down rivers and streams, sometimes scattering across hundreds of square kilometers and into the sea. These diamonds may be very close to the surface, leading to a frenzied diamond rush when they are first discovered, or they may be under 10 or 20 meters of overburden, the detritus of 100,000 millennia. Regardless, they do not require the capital-intensive investment of an undisturbed kimberlite pipe. Mining these diamonds requires little more than ingenuity and muscle.

As the story unfolds, it will become clear that the difference between diamonds from the two sources – kimberlite and alluvial – has made all the difference in both regulation and the potential for diamond-related development. While there is no guarantee that a kimberlite mine will be well managed or that it will become an engine of development, the possibility and examples exist. In the case of alluvial diamonds, the opposite has been true. Inexpensive to mine and almost impossible to police, alluvial diamonds are accessible to very poor people, but they have also been a source of chaos, social upheaval, and political corruption. And they were at the very epicenter of Africa's worst post-Cold War conflicts.

Crystallography and history

The crystallography of diamonds has also shaped the political economy of the industry and efforts at regulation. Three-quarters of the world's diamonds are so small, so badly formed or so poorly colored that they are relegated to industrial use as abrasives for metalworking and drilling. The value of these diamonds – worth as little as 30 cents, and usually not more than $10 a carat – is based on one of the diamond's primary features, its hardness. And with good reason: diamond is the hardest known natural material on earth. Its name, appropriately, derives from the Greek *adámas*, "unbreakable," a root still used in English for "adamant," meaning firm and unyielding.

The purest gem diamond is made up of nothing but carbon, crystallized into different forms: the eight-faced octahedron, the six-faced cube, the twelve-faced dodecahedron, the four-faced tetrahedron, and others. When they were being formed, some diamonds picked up impurities – nitrogen, hydrogen, boron – and these affect both color and luminescence. This in turn may provide a clue as to where the diamond was mined. "Blue" diamonds are found mainly in India, while most Canadian diamonds run from clear to a very light yellow. But yellow and blue diamonds are found elsewhere and it is virtually impossible, even for the most trained eye, to identify the geographical source of a polished diamond. It is more possible to make an educated guess about the origin of rough diamonds if they are in a group originating from the same source. It is likely, however, that an expert would be more confident in saying what they are not than what they are. Sierra Leone, for example, has a high-quality "run of mine average." Looking at a parcel of unlabeled Sierra Leonean diamonds, it would be easy enough to say that they are not from Côte d'Ivoire, where the average quality is significantly lower. Most of the diamonds mined at Marange in Zimbabwe are coarse and low in quality, with distinctive brown and black coloration. In a parcel, they are easily enough identified as Zimbabwean. But Marange's better-quality gems, while still of green and brownish hues, are similar to stones found in most other diamond mines around the world.

The fact is that, once mixed with others – and often before, in most cases – it is impossible for even the most educated and experienced eye to know where a rough diamond originated. This feature, historically of little interest beyond the mining and investment community, became extremely important during the 1990s and afterwards in the efforts to track and trace conflict diamonds, and to create a global regulatory system.

But the story is getting ahead of itself. What distinguished diamonds in antiquity was their hardness, and only the best-shaped crystals – uncut and unpolished – were used for jewelry. There are several references to diamonds in the Bible, the oldest in the Book of Exodus, thought to have been written in the sixth century BCE. The "breastplate of judgement" is said to contain an emerald, a sapphire, and a diamond (Exodus 28:18), and diamonds appear again in Ezekiel, Jeremiah, and Zechariah as metaphors for the hardness needed to cut stone, or the hardness of human hearts. In Roman times, Pliny the Elder described the properties and value of diamonds in his *Natural History*, saying, "The substance that possesses the greatest value, not only among the precious stones, but of all human possessions, is *adamas*; a mineral which, for a long time, was known to kings only, and to very few of them."[1]

Diamonds, like other gemstones, were said to have magical powers, assisting in childbirth, warding off epilepsy and the evil eye, detecting poison. It was not until the fourteenth century that they were first cut and polished, revealing a more authentic feature, one that made them – then and now – the most prized of jewels: the way in which white light is refracted off the facets to create a spectrum of colors known as "fire."

Pliny and the Biblical writers before him were referring to Indian diamonds, the only known source until the eighteenth century. Arab traders were the purveyors of these diamonds for centuries, but it was a French traveler and connoisseur, Jean-Baptiste Tavernier, who brought the most spectacular Indian diamonds to the attention of European royalty. Between 1630 and 1680, Tavernier visited and documented the mines of Soumelpur, Raolconda and Kollur, the source of the 108-carat Koh-i-noor. The Koh-i-noor, "Mountain of Light," is now set in a crown that spends most of its time in the Tower of London with the other British Crown Jewels.[2] Indian mines also produced the 280-carat Great Mogul, the 140-carat Regent Diamond, and the Tavernier Blue. Tavernier brought his 112-carat namesake back to France where he sold it to Louis XIV. Later it was cut into three stones, the largest of them the fabled Hope Diamond, now part of a necklace residing in a sealed glass vault at the Smithsonian Museum of Natural History in Washington.

Indian production declined rapidly after 1700, replaced by diamonds from an unexpected new source, Brazil. In 1725, deposits of rough diamonds were discovered in the eastern province of Minas Gerais, in an area where the primary diamond settlement grew into a town that was named for its fabled output: Diamantina. More diamonds were discovered in Bahia province and Mato Grasso – alluvial diamonds, easily accessible to diggers known in Portuguese as *garimpeiros,* meaning "prospectors." During the eighteenth century, Brazilian diamond production rose to 50,000 carats a year, flooding the world market and serving to cheapen a once-rare commodity. Prices became so depressed, in fact, that production dropped and failed to recover for almost a century. It was not until the late 1800s that Brazilian production recovered, rising to 200,000 carats a year or more. But by then Brazil had been eclipsed by South Africa, where discoveries in the 1860s set in motion a chain of events that would change the diamond industry for all time.

Some stories have the first South African diamond discovered by a Griqua boy near the Vaal River in 1859. A more common tale makes 15-year-old Erasmus Jacobs the hero, playing a game of "Five Stone" near Hopetown, 20 miles south of the Vaal on the Orange River, with a rock that turned out to be a 21-carat diamond. Certainly it was Jacobs' find, later cut into the 10.73-carat Eureka Diamond, that started one of the greatest diamond rushes of all time. By 1870, thousands of adventurers had crowded into the region and more than 10,000 claims had been sold. By the end of that year, an estimated £300,000-worth of rough diamonds had been found – no mean sum. But, given the number of people involved, gross incomes worked out to less than £60 apiece.[3] Part of the problem was that most of the newcomers had no idea where to look. The true geology of diamonds was unknown, and even when interest began to focus on the fields of a farm known as Vooruitzigt – "Foresight" – nobody had the actual foresight to see what was to come. The farmers who owned Vooruitzigt, two brothers named Johannes and Diederik De Beer, thought they had done well out of diamonds, selling their land for 6,000 guineas and leaving their name behind for posterity as a synonym for a gem they never mined and never owned.

By the spring of 1872 there were 50,000 people in the De Beer area, and at New Rush, not far away, there were so many diggers moving so much earth that a vast crater was beginning to form. They called it the Big Hole. A year later things had changed again. New Rush was renamed for the Colonial Secretary, John Wodehouse, the First Earl of Kimberley. New laws were passed to exclude "natives" and "coloured persons" from mining, instituting a penalty of 50 lashes to any found in possession of a diamond for which they "could not satisfactorily account."[4] The yellow earth of the Big Hole was giving way to a harder blue clay that would later be called "kimberlite," and a tall, delicate 20-year-old English *parvenu* named Cecil Rhodes was haunting the area, buying up as many claims as he could lay his hands on.

Geography

Diamond discoveries in other parts of the world are as much tales of accident as of long, painstaking exploration. A railway worker named Zacharias Lewala in German South West Africa – present-day Namibia – discovered a diamond in 1908 while he was shovelling sand away from the track. A frenzied rush into the desert ensued and some 7 million carats of high-quality diamonds were dug up in the six years before World War I broke out and the colony was wrested from German control.

A search for diamonds began in the Belgian Congo in 1906 under the auspices of the rapacious King Leopold II, and success was not long in coming. The first finds were on a tributary of the Kasai River called the Tshikapa, not far from the Angolan border. Later discoveries farther north at Mbuji-Mayi would soon turn the Congo into a major diamond-producing country. Although its diamonds were relatively low in quality, through much of the twentieth century Congo was the largest producer in Africa by volume, a distinction it retains today, only surpassed in some years by Botswana.

The Congolese discoveries sparked exploration in Portuguese Angola where diamonds were discovered in the northeastern Lunda province in 1912. Angola would prove to be a rich source of both kimberlite and alluvial diamonds, resulting over time in great wealth for some and great suffering for millions. The first West African finds did not occur until the 1930s, the richest in the eastern region of Sierra Leone – first alluvial and later kimberlite. Lesser pockets of alluvial diamonds would be located to the north and east in Guinea, Liberia, Côte d'Ivoire, and Ghana. In 1940, a Canadian geologist working in Northern Tanganyika, John Williamson, discovered one of the largest kimberlite pipes of all time. The Williamson mine would be a one-off for what is today Tanzania, low in gem-quality production, but important during World War II and since for its industrial stones.

The greatest and richest discoveries, however, were yet to come: Russia, Botswana, and Canada. Alluvial diamonds had been found in Russia's far north for generations but it was not until the late 1940s that exploration began in earnest. The Russian diamond legend is not unlike the tales of Erasmus Jacobs and Zacharias Lewala. Larissa Popugayeva was a geologist who knew what she was looking for, but luck was as important in her case as science. It is said that, while prospecting in a forest, she spotted a fox with the fur of its stomach stained a tell-tale kimberlite blue. Tracking the animal to its lair, she discovered that it had made its den in what turned out to be a kimberlitic mother lode. Today Russia produces more diamonds by volume than any other country in the world, second in value only to Botswana.

The Botswana story is surely the most fabulous of all, and, where diamonds are concerned, Botswana is without doubt the most fortunate of countries. A remote British protectorate beset with drought and desertification, Botswana – then known as Bechuanaland – had a population of barely a million people at independence in 1966, and a per capita gross domestic product (GDP) of only $70. One of the poorest countries on earth, it faced a questionable future. Within a year, however, everything had changed with the discovery of the AK1 kimberlite pipe, setting Botswana on a completely new course. AK1 became the Orapa Mine, capable of producing 2.5 million carats a year at its height. Nearby Letlhakane began production in 1977 and the Jwaneng Mine near the South African border opened in 1982, becoming the richest, in terms of the value of its production, in the world. In 2012, Botswana exported $4 billion worth of diamonds, slightly more by value than runner-up Russia, and exactly twice as much as third-place Canada.[5]

Smaller deposits of diamonds have been found in other countries of Africa, and in Guyana and Venezuela. For many years, Australia was the world's largest diamond producer by volume, although the quality of diamonds, primarily of industrial interest, was low. Coincidentally, the Argyle Mine is located in a district of Western Australia known as "The Kimberley," named for the same Colonial Secretary as its namesake in South Africa. At its peak in 1990, the Argyle Mine produced 33 million carats, about one-third of combined global production at the time.

One of the greatest diamond rushes of the past 100 years took place in 2006 at Marange in the eastern part of Zimbabwe. Much of Zimbabwe lies on what is known as the "Zimbabwe Archaean Craton," an ancient crystalline basement rock that is conducive to kimberlite deposits. Theoretically, it is possible that diamonds could be found anywhere in Zimbabwe, and three kimberlite pipes at Murowa in the south central part of the country had already led to a small mining operation there in 2004. But the Marange find was a different order of magnitude. The first diamonds, all alluvial, were discovered in September of 2006, and by December an estimated 10,000 people had flooded into the area, hoping to alleviate the dire poverty into which they and the rest of the country had been plunged. The story will return to Zimbabwe – the murders, the corruption, the human rights abuse, and the confusion into which it threw global regulatory efforts. But as an indication of the change that Marange wrought, Zimbabwean export numbers tell a tale. In 2004, Zimbabwe exported a total of 18,000 carats in rough diamonds. By 2012 the total had risen by a factor of more than 800 to 14.9 million carats.[6]

CHAPTER TWO

Supply and Demand – The Business of Diamonds

But first, some theory

The high price of diamonds not only defies the logic of supply and demand, it also defies the labor theory of value and most other standard economic explanations. Industrial diamonds have utility as an abrasive and in cutting and drilling, although this was not fully recognized until the 1930s. For industrial diamonds, the law of supply and demand then became obvious. But so low is their value that some mining firms actually "throw them back" as being unworthy of the effort in taking them to market. Gem diamonds, however, have only one apparent purpose – as jewelry – and given the tons of them that are mined each year, one must ask why supply has not driven the price down as more and more enter the market. If the answer has to do with demand, what is it about a diamond that keeps the price up regardless of how many are produced in a year?

Although Adam Smith had answers for a lot of questions, he had none for this one when, in 1776, he compared the usefulness of water – considerable – with the usefulness of a diamond – "scarce anything":

> The word *value* [he wrote] . . . has two different meanings, and sometimes expresses the utility of some particular object, and sometimes the power of purchasing other goods which the possession of that object conveys. Value may mean either value in use or value in exchange . . . The things

> which have the greatest value in use have frequently little or no value in exchange; and on the contrary, those which have the greatest value in exchange have frequently little or no value in use. Nothing is more useful than water: but it will purchase scarce any thing; scarce any thing can be had in exchange for it. A *diamond*, on the contrary, has scarce any value in use; but a very great quantity of other goods may frequently be had in exchange for it.[1]

Smith had stated what became known as the "paradox of value," and sometimes as the "diamond–water paradox."

Later economists argued that Smith had missed the idea of *marginal utility*, which posits that the value of a thing depends on its value to the individual. A farmer's first sack of grain will be used to make bread for his family's survival. The second will be for more bread, required for the strength to work. The third sack will go to the farm animals, the fourth to make whiskey, and the fifth to feed the farmer's pet pigeons. A shortfall in grain will not lead to a reduction in all five uses. The first to suffer will be the pigeons, and the last the family. The marginal value of the grain changes with the amount available.

That, of course, sounds like a variation on supply and demand rather than an explanation for why diamonds – readily available – are so costly. It does explain why, however, the *labor theory of value* – goods acquire their value because of the amount of labor used in producing them – does not apply in the case of diamonds. Or grain.

But grain, especially that first sack, does have genuine utility. It could be argued that a gem diamond does not. Enter the *subjective theory of value*. Here the value of goods and services resides not so much in their own intrinsic utility, as in the subjective value placed on them by the consumer. The consumer will place less and less value on each successive sack of grain, so marginal utility is still in play, but on top of that, each individual makes different subjective choices, seeking not only to maximize his or her monetary gain, but to satisfy needs for psychological and subjective gain as well.

That helps to explain why a painting by Monet is valued more highly than a poster of the same painting, or more highly than a painting by a lesser-known artist. Supply may be part of the equation – after all, there are not very many original Monets available. But subjective demand has a lot to do with it, and there was a time when demand for his paintings was so weak that Claude Monet could barely afford a loaf of bread, much less that first sack of grain. Despite the high price of a Monet today, there are many people – in Mongolia, say – who would place almost no value on one of his paintings, just as African diamond diggers – who may know what price to ask for a diamond – place no value on the ownership of one themselves.

The four C's

This is not to say that supply and demand have no role in the diamond industry; they most certainly do. It is only to say that other factors are often at work, not least the eye of the beholder. That aside, there *are* objective elements that help to create scarcity, demand, and objective value in a diamond. Quality is a major factor in the pricing of most goods, diamonds no less. And here quality is measured in what are known as the "Four C's": cut, clarity, color, and carats.

The "cut" of a finished diamond is all-important. The cut is different from the shape of the stone. Shape will be determined in part by the shape of the crystal, in part by the market. Some 75 percent of the diamonds produced today are what is known as the round brilliant shape, with 33 facets on the top portion or "crown," and 25 facets on the bottom portion or "pavilion." There are other shapes: the princess, cushion, oval, pear, heart shape and more, but the quality of the *cut* will always be the deciding factor in pricing the stone.

The popularity of the round shape derives from the fact that if a stone has been well processed, if the proportions and angles and finish are right, these will show off the diamond's "fire" and luminescence to best advantage. In the "ideal cut," there is a common understanding of what the proportions must be – the heights of the crown and pavilion, the stone's diameter and the size of the top or "table." If the diamond is too squat or too tall, light will not reflect properly off its facets, and will instead "leak" out of the diamond. The same will happen if the angles are wrong. There is a hierarchy of cuts: "ideal" or "excellent," "very good," "good," "fair," and "poor," each with its own specific meaning.

Each diamond is also graded for "clarity." A diamond may have a variety of flaws: surface flaws such as cracks, scratches and chips, or internal flaws such as air bubbles or other minerals. Known as "inclusions," these affect the value of a diamond, giving it a "flawless" grade (F), "internally flawless" (IF), "very very slightly included" (VVS1 and VVS2), "very slightly included" (VS1 and VS2), through to three grades of "included" (I 1, I 2, and I 3) – diamonds with heavy blemishes visible to the naked eye.

Color is important as well. Absolutely colorless stones, rated "D" by the Gemological Institute of America, emit the best sparkle and fire. An "E" stone will have very slight traces of light yellow or brownish color; an "F" stone will have more, on a scale continuing on down to "Z." Fancy-color diamonds are different: these may be any color from bright yellow to blue or green, and are more valuable because of it.

Finally, the carat weight of a diamond is obviously of great importance. Not to be confused with "karat," the method used for determining the purity of gold, a diamond "carat" represents 200 milligrams or 0.2 grams.

The variation in price can be enormous between two diamonds of the same shape and size. On a particular day in October 2012, two "very good" round cut diamonds of 1.02 and 1.04 carats were offered on eBay for $2,200 and $4,884 respectively. The first was graded "Q" in color, the second "H." The first had I 1 clarity, the second SI 1. Had the clarity been VS1 in the latter, it might have sold for an extra $2,000. Retain the SI 1 clarity but double the size to 2 carats in an "F" color and you would have had to pay almost $14,000. Retain the 2-carat weight, stay with the "F" color but move to VVS2 quality – very very slightly included – and you would have been looking at a price tag of more than $35,000.

Truly exceptional diamonds fetch astronomical prices. The Archduke Joseph Diamond, named for its one-time Habsburg owner, was sold at auction by Christie's in 2012 for a whopping $21.5 million, a new record for a colorless diamond. The Archduke Joseph is a D-IF – clear and internally flawless – weighing in at 76 carats. The price represents approximately $283,000 per carat.

Controlling the supply

There must be an estimate somewhere of the volume of diamonds produced since Jean-Baptiste Tavernier returned from his first trip to India. It would, at best, be speculative, but the quantity would be high. Historical data on diamond production[2] are murky for all manner of reasons – commercial confidentiality, security issues, and illicit behavior among them – but reasonably good current records exist, and approximate volumes are known. Between 2005 and 2011, 1.053 billion carats were produced, or about 210.6 metric tons, for an average of 30 metric tons a year.[3] If 75 percent of the total was industrial, then the average annual production of gem diamonds was approximately 7.5 metric tons, the weight of about six Toyota Corollas. This means that enough gem-quality diamonds were produced to give 1 carat of rough to every man, woman, and child in the United Kingdom, Germany, France, Spain, and Portugal *every year*.

Diamond mining has been going on for a long time, and diamonds do not wear out, fade, or degrade. *Diamonds are forever.* So Adam Smith would probably still have trouble with the price tags, even if someone were to explain to him the theory of subjective value and the difference between a D Flawless and an H VS1.

The Colossus of Africa

One of those who could perhaps have done it was Cecil Rhodes, although Rhodes had a more elemental idea about diamonds – a kind of Tony Soprano approach: control everything, get rid of the opposition, restrict the supply, and force prices up. Rhodes, the son of a Hertfordshire vicar, went to South Africa in 1870 at the age of 17. He had a weak heart and a collapsed lung, and it was thought that a better climate in the company of his older brother Herbert, a successful cotton farmer, would change his prospects. It did, but not because of the cotton. Herbert had purchased three diamond claims on land lying between the De Beers mine and New Rush, and soon young Cecil was scrabbling in the diggings, some of them 50 feet deep. In a letter to his mother, he wrote, "There are constantly mules, carts and all going head over heels into the mines below, as there are no rails or anything on either side of the road, nothing but one great chasm below."[4]

The Rhodes claims were good ones, and Cecil began to buy up neighboring operations. By 1873, that was not a problem – many miners were eager to sell. The flood of South African diamonds had seriously deflated European prices, and a global economic depression added to the industry's woes. In addition, many miners had reached a different type of rock bottom, believing that their claims were drying up as the yellow diamondiferous soil they knew gave way to a harder blue rock. To compound the problem, flooding became an issue as the mines probed deeper into the earth. In the flooding, Rhodes saw another opportunity. He bought the only steam pump in South Africa, hauling it across the veldt to Kimberley where his first monopoly – water pumping – added to the small fortune he had begun to accumulate.

Rhodes understood power and influence, running for office in 1880 and winning election to the parliament of the Cape Colony. By then it was clear to Rhodes – now 27 – that something had to be done to rein in the physical and economic chaos of the South African diamond industry. It was obvious that there was an almost unlimited supply of diamonds, and if they were ever to regain and hold their value, steps had to be taken. That year, Rhodes and five partners created the De Beers Mining Company Limited, with share capital of £200,000. Their aim, simply stated, was to regulate the production and marketing of diamonds, an audacious objective – not least because among the six partners they owned only 90 of the 622 registered claims in the De Beers mining area. But it was a start.

Their most serious competitors were the Kimberley Central Diamond Mining Company and the Compagnie Française des Diamants du Cap, known as the "French Company." Kimberley Central was controlled by a former English dancehall performer, one Barney Barnato, who – like Rhodes – understood that consolidation and control were essential to the wider plan they each had for diamonds. But neither wanted to do it together. In an effort to gain the upper hand, Rhodes bought as many shares as he could in any company on offer. Then, in 1887, with Rothschild backing, he offered an unprecedented £1.4 million for the French Company. Barnato countered with an offer of £1.75 million. Loath to bid the price higher, Rhodes took a new approach. He went to Barnato with an offer: if Barnato would withdraw his bid, Rhodes would buy the French Company and resell it to Barnato in return for £300,000 and a 20 percent share in Barnato' s Kimberley Central.

Barnato, thinking he had made the best of the bargain and that he could limit Rhodes' overweening ambition, seriously underestimated his opponent. After the deal was concluded, Rhodes went back to his European bankers and persuaded them that, without full centralized control, diamonds would continue to flood the market at bargain-basement prices. His plan was to buy up as many of Barnato's Kimberley Central shares as it would take to gain control. There are differing versions of what happened next. The most dramatic is that Rhodes flooded the market with diamonds, depressing the price not only of diamonds, but of Kimberley Central shares, which he then bought for pennies on the pound. There are other versions of the "war" between Barnato and Rhodes, but one thing is clear: by 1888, Rhodes and his colleagues had the number of Kimberley Central shares they needed for takeover and amalgamation. They liquidated Kimberley Central, selling its assets to De Beers for the unprecedented sum of £5,338,650. The cheque, the largest ever written at that time – and worth 50 times that amount at today's prices – remains on display in the old De Beers boardroom in Kimberley.

The impact was immediate. As partners and shareholders in the new De Beers Consolidated Mines, Rhodes, Barnato, and their colleagues became rich beyond imagination. Diamond production was immediately curtailed and the price of rough increased by 50 percent. A quarter of the white labor force was fired; and an estimated half of the black work force – which had provided the bulk of the labor that made Rhodes and his colleagues so wealthy – was simply sent away. "I prefer land to niggers," Rhodes once said, no doubt thinking of the vast area to the north that would soon bear his name.[5] From now on, Rhodes would require police protection, but at last he controlled more than 90 percent of the world's diamond production.

In the two years following the takeover, Rhodes was a busy man. He bought up two small remaining diamond pipes at Dutoitspan and Bultfontein and established control over European diamond marketing by selling the entire De Beers production through a syndicate of ten London diamond merchants. In 1889 he created the British South Africa Company with the aim of annexing large parts of the continent, and the following year, during an influence-peddling scandal, he used his growing political skills to manipulate his way into the job of Prime Minister of the Cape Colony.

Diamonds, which might have been a source of economic growth and development for the original inhabitants of South Africa, now became an instrument of their oppression. In the years ahead, Rhodes pushed the British Empire into Matabeleland and farther north to areas that for a while would bear his name – Northern Rhodesia and Southern Rhodesia. And he began to frame laws and consolidate attitudes that would haunt Southern Africa for 100 years to come: "There must be class legislation . . . there must be Pass Laws and Peace Preservation acts . . . we have got to treat the natives where they are in a state of barbarism, in a different way to ourselves. We are to be the lords over them."[6]

And lords over the world's diamonds. By controlling the world's supply of diamonds and narrowing the London funnel through which they flowed, Rhodes had created a model that would shape the industry too for more than a century, surviving a myriad of commercial challenges, American anti-trust laws, the Depression, two world wars, the end of colonialism, apartheid and most of what Rhodes stood for. Mark Twain visited Kimberley in 1897 and witnessed the wealth, the distressed African population, and the diamonds. "The De Beers concern treats 8,000 carloads," he wrote, "about 6,000 tons of blue rock per day, and the result is three pounds of diamonds. Value, uncut, $50,000 to $70,000. After cutting, they will weigh considerably less than a pound, but will be worth four or five times as much as they were before."

Twain marveled at Rhodes' crimes and his ability to get away with them: "He raids and robs and slays and enslaves the Matabele and gets worlds of Charter-Christian applause for it . . . I admire him, I frankly confess it; and when his time comes I shall buy a piece of the rope for a keepsake."[7]

There would be no rope, however. Rhodes would not hang; ill-health and a bad heart soon caught up with the "Colossus of Africa." One of the wealthiest men in the world, he died in 1902, not yet 50. In an obituary, the French newspaper *Le Temps* said that Rhodes "lived only for his schemes and enjoyed life only as a cannon ball enjoys space, traveling to its aim blindly and spreading ruin on its way. He was a great man, no doubt a man who rendered immense services to his country, but humanity is not much indebted to him."[8]

What remain are the scholarship fund he endowed at Oxford, a memorial at Devil's Peak in Cape Town, and De Beers.

The Oppenheimer Factor

Rhodes was in his grave barely a year when the first challenge to his great monopoly arose. A building contractor named Thomas Cullinan opened the Premier Mine 40 kilometers northeast of Pretoria. By 1904 the mine was a going concern and a year later it produced one of the world's most spectacular stones, a 3,025-carat colorless, single-crystal diamond. The principal stone later produced from this diamond, the 530-carat Cullinan 1, the "Great Star of Africa," is now set in the Scepter of the British Crown Jewels.

The Premier Mine posed a major threat to the concept Rhodes had of the industry. De Beers had attempted to purchase the Premier operation even before its luminous find, but to no avail. Production at the Premier Mine continued apace, growing to more than 2 million carats a year by 1912. For De Beers, there was an additional problem across the border in German South West Africa. The discovery of diamonds there in 1908 had led to the production of an estimated 7 million carats by the summer of 1914. With diamonds now flooding the market, prices plummeted, validating Rhodes' thesis on the need for control. De Beers now represented less than half of the world's production, and it seemed that Rhodes' diamond empire, 15 years in the creation, was in a state of freefall.

Two things intervened. One was the outbreak of war in August 1914. The Premier Mine, already suffering from low prices and labor unrest, closed. South African troops attacked German South West Africa less than a month after European hostilities began, and soon controlled the entire colony. Its major competitors now in limbo, De Beers could take a breath as the war played itself out.

De Beers' revival would spring from an unlikely source – a young German diamond buyer who arrived in Kimberley at the age of 22, only a few weeks after the death of Cecil Rhodes. Ernest Oppenheimer had moved from Germany to London when he was 16 and worked as a sorter with Anton Dunkelsbuhler, one of ten buyers in the diamond syndicate. But the magnet for young Ernest was South Africa where, in due course, he set up shop on behalf of his London employer. There he saw at first hand the damage that the flood of Premier and South West African diamonds was doing to prices, and in 1910 he wrote, "The only way to increase the value of diamonds is to make them scarce, that is, to reduce production."[9] Like Rhodes, Oppenheimer had an eye for politics and was elected to the Kimberley town council in 1908. While the South African diamond business floundered, he pursued a political career, becoming Mayor of Kimberley in 1912. He was 32 now, and a successful politician, but it wouldn't last. The outbreak of war effectively shut down the diamond business, and a mayor with a German background and a Teutonic name had no place in Kimberley.

Fortunes change, however, especially for those with a good eye for opportunity. Both his German background and his newly acquired British citizenship proved fortuitous. With growing British and South African clamor for the expropriation of enemy assets, Oppenheimer created a company in which German investors could protect or offload their South West African assets. Looking for a high-sounding name that would reflect the presence of the American money he had attracted, Oppenheimer called his new company the Anglo American Corporation of South Africa Ltd. The Germans received shares in the new company or cash, and Anglo American, with Ernest Oppenheimer as Chairman and Managing Director, soon owned most of the fabulously wealthy diamond operations in South West Africa's *Sperrgebiet* or "forbidden zone." Oppenheimer reorganized these operations into a company he called Consolidated Diamond Mines. He invested in gold and other businesses, was knighted in 1921, and returned to politics as the Member of Parliament for Kimberley in 1924. He acquired rights to new diamond finds in Angola and the Belgian Congo. He had *arrived*.

But he did not control. Oppenheimer had long coveted a position on the De Beers Board of Directors, but he had more than once been rebuffed. Now, in order to consolidate his position in the industry, he partnered with an older man, Solly Joel, a flamboyant "Randlord" who had become wealthy buying diamond mines that others thought depleted. Joel was also, coincidentally, a nephew of Barney Barnato, who, after years of mental instability, had ended his own life in 1897. Joel now ran the considerable Barnato business empire. He also owned a large chunk of De Beers and was a member of the diamond syndicate. Perhaps when he joined with Oppenheimer, Joel was thinking about how Rhodes had got the better of his uncle in 1888. Perhaps it was simply the challenge, or just good business sense. Whatever it was, they made a formidable team when Oppenheimer created a new diamond syndicate to rival De Beers, with himself at the helm. Not only did he now control 49 percent of the world's diamond production, he and Joel were quietly buying up as much De Beers stock as they could lay their hands on. By 1926, De Beers could no longer keep him out, and Ernest Oppenheimer was finally appointed to the Board of Directors. Three years later, in 1929 – now the most powerful man in the diamond industry – he took over the De Beers chairmanship. Merging Consolidated Diamond Mines into De Beers, he brought his Anglo American company into the picture as the controling shareholder.

The Cartel

For De Beers and the diamond industry, Oppenheimer's appointment happened not a moment too soon. In fact, coming only a few weeks after "Black Tuesday" and the Wall Street crash of 1929, his arrival was almost a moment too late. The only person with the fortitude and the tenacity to resuscitate Cecil Rhodes' dream during the economic disaster now consuming the world was Ernest Oppenheimer. But Oppenheimer differed from Rhodes. He had begun to see that Rhodes' idea of controlling *production* would always be threatened by new diamond finds and other investors in the years ahead. Although De Beers would attempt as much as possible to control global production, it would also do something else: it would control global *distribution*. One of Oppenheimer's first moves was to scrap the syndicate, replacing it with something called the Diamond Corporation. Oppenheimer put his brother Otto in charge, with a mandate to buy diamonds from inside and outside the De Beers group, and to release them only as fast as the market could absorb them. In addition, De Beers would continue to seek out and exploit new diamond mining areas. And if others brought diamond mines on-stream, De Beers would buy them or their production, making them offers they couldn't refuse.

Gathering all of the world's rough diamonds into the Diamond Corporation, Oppenheimer then developed a unique marketing approach through a firm he called the Diamond Trading Company (DTC). The DTC would sell diamonds only to a limited number of pre-approved buyers at periodic events that became known as "sights." The buyers were called "sightholders" and they would be invited to London to examine boxes of diamonds prepared for their individual needs. The diamonds were offered on a take-it-or-leave-it basis and the price was not negotiable. De Beers would try to meet the specific needs of each company, but in order to offload stones that nobody wanted, each parcel would inevitably contain "extras." The sightholder could, of course refuse to buy the parcel and hope for better next time. But in the meantime, he had no diamonds, and there might very well not be a next time. "Single-channel marketing" was the term for the overall system – a polite euphemism for "monopoly."

It was this single-channel marketing, the "cartel" and its monopolistic behavior that saved the diamond industry during the worst of the 1930s, and it would continue to protect the industry against the vicissitudes of war, politics, ideology and sharp economic fluctuation in the years ahead. De Beers persuaded the Soviet Union, for example, that partnering with an almost perfect exemplar of what Marxian economists called "monopoly capitalism" made sense – following the discovery of Yakutia's vast diamond resources in the 1950s. And through a network of subsidiaries that offered plausible deniability, this quintessentially South African company was able to conduct a thriving business in post-colonial Africa with governments that were unalterably opposed to apartheid. The greatest challenge to the monopoly came from a rather unexpected source, a country where almost half the world's gem diamonds are sold: the United States.

The 1890 Sherman Anti-Trust Act and similar laws were designed to break up and prevent the recurrence of monopolistic behavior among giant railroad companies, banking and insurance firms, oil companies, and others. In a series of judgements over the years, the US Supreme Court deemed that monopolistic behavior was a "restraint of trade," a detriment to business that harmed the consumer and violated standards of ethical practices through price-fixing and predatory behavior. A 1972 US Supreme Court decision described anti-trust laws as: "the Magna Carta of free enterprise. They are as important to the preservation of economic freedom and our free enterprise system as the Bill of Rights is to the protection of our fundamental freedoms."[10]

Given the almost religious zeal with which anti-trust legislation was enforced in the United States, it was only a matter of time before De Beers would run afoul of it. The US Justice Department filed its first case against De Beers in 1945. The effort failed, largely because De Beers operated entirely outside American jurisdiction. American buyers traveled to Europe to buy from De Beers; De Beers did not operate in the United States. The second case centered on a complex web of companies and agreements in the US industrial diamond sector. During the 1970s, the Justice Department alleged that, through subsidiaries and joint investments, De Beers had engaged with a US business, the Christensen Diamond Product Company, in a "conspiracy . . . to suppress competition . . . and to fix, maintain and stabilize world market prices for such diamonds."[11]

De Beers managed to evade prosecution by shedding Christensen stock, but the Justice Department was back again with a new suit in 1994 against both General Electric and De Beers in another industrial diamond price-fixing charge. The case against General Electric was dismissed, but De Beers never showed up in court to defend itself, remaining for a decade under a legal cloud in the United States. The company was unrepentant. In a 1999 speech, Nicky Oppenheimer, grandson of Sir Ernest, introduced himself to a Harvard Business School audience this way:

> I am chairman of De Beers, a company that likes to think of itself as the world's best known and longest running monopoly. We set out, as a matter of policy, to break the commandments of Mr. Sherman. We make no pretence that we are not seeking to manage the diamond market, to control supply, to manage prices and to act collusively with our partners in the business.[12]

He defended "single-channel marketing" in two ways. First, he said, by controlling the price De Beers was protecting the consumer, ensuring that the purchase retained its value. He said that any company dealing in a total luxury "cannot behave as an evil monopoly exploiting the masses, because at the end of the day they do not have a compelling need to purchase." Secondly, he said, "single channel marketing has exercised an extraordinary beneficial influence upon the whole of the diamond industry, and particularly many of the economies of Africa."

But the times were changing. The claim of benefits to Africa was about to come undone in the blood diamond wars. And De Beers' share of the market was dropping precipitously. Between the time when Nicky Oppenheimer so enthusiastically defended the company's monopolistic behavior and

Table 2.1 Volume and value of rough diamond production, major diamond-producing countries, 2012

	Production		
Country	Volume, carats	Value, US$	US$ per carat
Angola	8,330,995	1,110,222,942	133.26
Australia	9,180,923	269,419,306	29.35
Botswana	20,554,928	2,979,400,296	144.95
Canada	10,450,618	2,007,217,350	192.07
Congo, Democratic Rep.	21,524,266	183,135,861	8.51
Lesotho	478,926	301,452,475	629.43
Namibia	1,628,779	900,479,643	552.87
Russian Federation	34,927,650	2,873,728,990	82.28
South Africa	7,077,431	1,027,131,959	145.13
Zimbabwe	12,060,162	644,033,522	53.40

Source: Kimberley Process Rough Diamond Statistics, https://kimberleyprocessstatistics.org/public_statistics.

How to read these statistics: The key is the per-carat average. In first place, Russia produced 70 percent more diamonds by volume than runner-up Botswana, but the value of Botswana's total production was 3.7 percent higher than Russia's. Russia produced more than three times the Canadian volume of diamonds but by value they were worth only 43 percent more. The DRC produces 45 times more diamonds by volume than Lesotho, but the value of Lesotho's total production is almost double that of the DRC because of Lesotho's very high per-carat average. Zimbabwe was the world's fourth-largest producer by volume in 2012 but seventh in terms of value, while Australia, the seventh largest producer by volume, was the ninth largest by value.

2004, the company's share of the world's rough market plummeted from 80 percent to something in the neighborhood of 45 percent. Several things happened almost at once. Rio Tinto, owner of the Argyle Mine, Australia's largest, had already abandoned the cartel in 1996. Next, Canadian diamonds began to enter the market. BHP-Billiton, the biggest mining company in the world, decided to market the product of its lucrative Ekati Mine outside the cartel, as did Rio Tinto with Diavik, the second Canadian mine to begin production. De Beers would eventually edge into the Canadian market

Table 2.2 Volume and value of rough diamond imports and exports, major importing countries, 2012

Country	Import		Export	
	Volume, carats	Value, US$	Volume, carats	Value, US$
China, People's Republic	21,141,886	2,730,268,112	15,030,127	1,664,211,397
European Community	124,813,783	16,789,891,223	126,882,182	17,813,699,529
India	151,873,237	14,884,640,599	34,439,402	1,803,560,231
Israel	13,271,865	4,657,945,870	13,789,834	3,610,105,573
Switzerland	8,683,854	1,985,973,937	8,879,739	2,287,605,528
United Arab Emirates	59,744,046	4,564,338,478	60,446,766	6,821,264,076

Source: Kimberley Process Rough Diamond Statistics, https://kimberleyprocessstatistics.org/public_statistics.

Interpretation: China imported 21.1 million carats, processing roughly 6 million and re-exporting the rest. The European Community, Israel, and Switzerland exported around the same volume as they imported, demonstrating their roles primarily in diamond trading rather than processing. Indian re-exports represent only 22.6 percent of imports, demonstrating that most diamonds go to India for cutting and polishing. The anomaly in these statistics is the United Arab Emirates, where the numbers indicate a 49.4 percent mark-up on a volume of exports only 1.2 percent greater than imports. This could represent unusually high profit margins, or it might represent the "round-tripping" discussed in chapter 6. Or it might be something else entirely. All of these numbers are open to broad interpretation, as diamonds may pass through the same center more than once before they reach the place where they are cut and polished.

with its own mines at Snap Lake and Attawapiskat, but in the early years of the decade the company was well behind the Canadian eight-ball.

A third factor was the rise of a former De Beers sightholder, Lev Leviev, a canny Israeli millionaire hailing originally from Uzbekistan. Leviev, close to Vladimir Putin, was able to horn into De Beers' arrangements with Alrosa, the Russian diamond mining giant. He made deals in Angola that sideswiped De Beers' traditional toe-hold there. And he appealed to other African governments' interest in adding local value by setting up cutting and polishing operations in Namibia, Botswana, and Angola. In a few short years, Leviev had created a vast and vertically integrated diamond empire that rivaled the achievements of Ernest Oppenheimer, threatening the very essence of what Oppenheimer had created. Forbes called Leviev "the man who cracked the De Beers cartel," ranking him at No. 210 among the world's richest men, with an estimated net worth of $4.1 billion.

Diamonds may be forever, but just as things had changed for De Beers, they would soon change for Leviev. He made some bad deals and some bad friends. By 2012 Forbes had him down to No. 764, with a significantly depleted net worth of $1.5 billion.

Meanwhile, De Beers' outstanding US court case was finally settled in 2004 when the company pled guilty and paid a $10 million fine, a small price in relation to the market that the company could now enter legally and openly. There would be other cases, class-action lawsuits brought by American companies, the largest of which resulted in a 2012 $300 million De Beers pay-out in settlement of longstanding claims of price-fixing and monopolistic behavior. A De Beers spokesman said at the time, "This normalizes business for De Beers in America. If you think about the past century, this is a milestone moment."[13]

It was, and, in a sense, it marked the end of an era. The DTC system had been refined over the years. The sights no longer took place exclusively in London. The shoeboxes crammed with envelopes full of rough diamonds had been replaced by tasteful yellow-and-black cases. The term "single-channel marketing" had been superseded by a "supplier of choice" strategy. And despite the fact that De Beers was still the largest player, it no longer controlled the industry as it once did. Ernest Oppenheimer might have lamented the fact that his grandson, Nicky, sold the family's 40 percent share in the company in 2012, but he would no doubt have been pleased that at least the basic elements he set in place remained, the price of diamonds was still under a modicum of control, and the new owner of De Beers was none other than the company he set up in 1917 to engineer the first takeover: Anglo American. He probably wouldn't have sneezed at the price either: $5.1 billion.

Downstream

Throughout the twentieth century, cutting and polishing was carried out by De Beers' sightholders, or by their customers, or by their customers' customers. The downstream business is as vast as it is diverse, and at its center lies Antwerp. Geopolitics and religious bigotry contributed to the rise of this small port city as the world's diamond capital. In ancient times, Indian diamonds passed through Jewish trading centers in the Middle East on their way to Europe, and later through Jewish trading houses in Portugal. Because of the restrictions imposed on Jews in Europe, cutting and polishing diamonds was one of the few crafts open to them. In the sixteenth century, Antwerp became a trading hub for sugar from Portuguese and Spanish colonies in the New World, attracting other forms of business and banking and generating a cosmopolitan outlook on the world. The Portuguese connection served to bring the diamond trade to Antwerp, and until recently the business was populated almost exclusively by Jewish artisans and traders.

Because of the concentration in Antwerp, the city has been associated with most of the major advancements in diamond technology. The unique characteristics of a diamond are not usually evident in its rough condition and it requires experience to identify the grain and to "cleave" or cut the crystal. Cutting is the first step, polishing the next. It was in Antwerp at the end of the fifteenth century that a Jewish diamond cutter named Lodewyk van Berken invented the "scaif," a polishing wheel impregnated with olive oil and diamond dust. By placing a diamond in a holder known as a "dop" and pressing it against a spinning scaif, a polisher could create new angles, greater symmetry and a better finish than had yet been seen. Antwerp became known for the quality of its output and, in addition to trade, soon became the world's preeminent center for cutting and polishing. The sawing of diamonds was invented here, as was the formula for the "ideal" proportions of a finished diamond. There were other reasons for Antwerp's diamond importance, not least the discovery of huge diamond deposits in the Belgian Congo and Ernest Oppenheimer's desire to maintain the very best relationship possible with the Belgian government.

During the mid twentieth century, other centers began to develop. To avoid high US import taxes on finished goods, some of the better diamonds were taken to New York for cutting and polishing. At the opposite end of the spectrum, some of the smallest stones were sent to India where low wages justified the effort to put a few facets on tiny chips that might add glitter to a piece of jewelry. An Israeli diamond industry began almost accidentally in 1939 with the arrival of two Jewish refugee cutters in the British Palestine Mandate. As German armies overran Belgium and the rest of Europe, other Jews moved to Palestine and more were trained. During the war years, De Beers shipped an estimated $100 million worth of diamonds to Palestine,[14] and although Antwerp began to recover during the late 1940s, an Israeli industry for "melee" – small diamonds of 0.18 carats or less – had been established.

Just as the Israeli government encouraged the growth of a domestic industry, the government of newly independent India did the same, first with Indian-mined diamonds and later with imported goods. The Indian effort began as a cottage industry, heavily reliant on semi-skilled workers, small operations and low-value diamonds. By 1971, Indian production had grown to what seemed like an impressive 530,000 carats.[15] Today it is many times that size. There are hundreds of thousands of cutters and polishers in India, at least half a million of them in Surat, a 4-hour train journey from Bombay. Surat, unknown to most of the world, is one of the fastest-growing cities on earth, and diamonds are at the heart of its expansion. In 2011 India imported over 132 million carats in rough diamonds,[16] and in 2011–12 – a bad year for the industry because of sluggish recovery from the recession – it exported $23 billion worth of polished goods. Polished diamonds were, in fact, India's biggest single export, accounting for 60 percent of the world's polished diamond supply by value and 85 percent by volume.

As in most things, China is a place to watch where diamonds are concerned. Production costs are higher than in India, but demand is growing and Chinese companies are starting to turn their attention towards suppliers of rough diamonds in Africa. Today's smallish Chinese cutting industry, with 20,000 to 30,000 workers, could become tomorrow's next big thing.

Antwerp, therefore, is no longer the heart of the world's cutting and polishing, but it remains the center of the global diamond trade. Dubai has also become an important diamond trading hub. But an estimated 70–80 percent of the world's diamonds still pass through Antwerp at least once on their way from one dealer or one country to another, and the city is alive with diamond traders, wholesalers, manufacturers, banking houses, insurance companies, gem labs, and security shipping firms. Orthodox Jews are still much in evidence, but there are also Indians now, along with Russians, Lebanese, Chinese, and many other nationalities. Amidst the kosher restaurants and butchers in the diamond district one can also find Indian restaurants and even an Indian supermarket. Turnover in Antwerp's diamond business in 2011 was estimated at $56 billion, no small amount.[17]

Creating demand, pushing preciousness

There cannot be very many people living within a mile of a jewelry shop who do not know the De Beers advertising slogan, "A diamond is forever." It's one of the advertising world's great and most durable slogans. For both endurance and fictional content, it's right up there with "Guinness is good for you" and "Wheaties: Breakfast of Champions." Diamonds are the hardest known mineral, but hit one with a hammer and you'll get an idea of how lame the "forever" part can be. The slogan is credited to an exhausted copy-writer working late one night in 1948 at the American advertising firm of N. W. Ayer & Son. The company was paid well for it, but the slogan sounds suspiciously like a line uttered by Lorelei Lee, a "professional lady" from Little Rock who first appeared in Anita Loos' best-selling 1925 novel *Gentlemen Prefer Blondes*: "Kissing your hand may make you feel very good, but a diamond bracelet lasts forever," says Lorelei.[18] She would say it again in a 1949 Broadway musical based on the book, and more famously in the 1953 Monroe film classic.

The interesting thing about the De Beers advertising slogan is not so much what it says about diamonds as what it neglects to say about De Beers. For the first 50 years the ad was in play, there was no such thing as a polished De Beers diamond. No jewelry shop anywhere sold a "De Beers diamond." Although most rough diamonds passed through De Beers' hands at some point in their journey from mine to showcase, De Beers did not cut and polish diamonds and had no retail operation at all. Its only business was rough diamonds and the advertising was purely generic. It was not until 2001 that De Beers entered the retail trade in partnership with the French luxury goods company Louis Vuitton Moet Hennessy (LVMH).

Historically, the idea of an engagement ring or a betrothal ring goes back to Roman times. The first known diamond engagement ring was reputedly commissioned for Mary of Burgundy by the Archduke Maximilian of Austria in 1477. By the late nineteenth century, the idea was common enough for Cecil Rhodes to begin pegging the volume of diamonds he would release onto the world market against the number of "licit relationships" – his odd expression for wedding engagements – that could be predicted in a given year. The number of diamond engagement rings grew, of course, with the greater volume and lower prices of South African diamonds. It was not until the 1930s, however, that De Beers began to think about actually creating demand.

By 1938, 75 percent of De Beers' diamonds were eventually set in American engagement rings, but the diamonds tended to be small and cheap. Harry Oppenheimer, son of Sir Ernest, and his successor as Chairman of De Beers after 1957, engaged N. W. Ayer & Son in a concerted campaign to change all that. The "forever" slogan would come later, but first the idea was to crank up media stories about diamonds and to create a new image for them. Early efforts at what is now called "product placement" saw diamonds becoming central to the plots of romantic films and radio plays; diamonds were advertised frequently in up-scale magazines; diamonds appeared conspicuously on the fingers of Hollywood screen stars; and positive "news" stories about diamonds became more frequent. Instead of cheap rings, young men were encouraged to think that an engagement ring should cost a month's salary, later doubled to two. *Reader's Digest* was used to advertise diamonds internationally and, after World War II, campaigns introduced the diamond engagement ring in Sweden and Germany. Japan, a country where diamond engagement rings were virtually unknown in 1968, underwent an advertising blitz, and by 1981, 60 percent of Japanese brides wore a diamond engagement ring.[19]

Marilyn Monroe made diamonds a girl's best friend, and diamonds became the *leitmotif* in countless films: *To Catch a Thief, Diamonds Are Forever, The Pink Panther, The Bank Job, A Fish Called Wanda,* and dozens more. In all of them, the allure and mystique of diamonds were front-row center. Diamond: the most precious and most desirable thing in the whole world.

"Ten-year anniversary rings" were invented and heavily promoted during the 1960s when De Beers had a sudden flood of cheap Soviet diamonds to dispose of, and the company later created "eternity" bands. Then, perhaps in view of a growing market among unmarried couples, came the "three-stone ring" – good for just about anything:

> All women would like a three stone ring from her partner [*sic*]. The three stone ring was designed by the De Beers Jewellery Company and became an instant classic, setting the jewellery industry on its ear. The three stones signify the message of love to the recipient. The first stone, often smaller, signifies past love. The middle and large stone represents present love and the smaller third stone is for future love.[20]

At the beginning of the twenty-first century, as De Beers began to lose its monopolistic control over the supply of diamonds, it escalated its efforts on the demand side, creating the "Forevermark" ("Each Forevermark diamond is inscribed with a promise, a promise that it is beautiful, rare and has been responsibly sourced"[21]). And it reduced its list of sightholders to favor those who "added value" to the industry through their own increased advertising. Gary Ralfe, De Beers' Managing Director between 1998 and 2006, hated the idea of anything to do with cheapness. At the time of De Beers' entry into the retail market and its partnership with LVMH, he said: "We want to see stores pushing the preciousness of diamonds rather than treating them as a commodity you can discount."[22]

Can preciousness adequately substitute for the production control that Cecil Rhodes and Ernest Oppenheimer saw as an essential feature of price protection? Or will the flood of diamonds and the onslaught of undisciplined freebooters kill the goose that over so many years laid so many golden eggs? Too many diamonds and too many players are what wrecked the Brazilian industry in the late 1700s and it did the same in South Africa in the 1870s. It happened again before World War I and at the beginning of the Depression.

In September 2012, news began to appear about a "long-suppressed" diamond find in Russia, at the Popigai crater in Siberia. "We are speaking about trillions of carats!" a breathless *Daily Mail* said, quoting a Russian official. The subject was a "62 MILE-WIDE asteroid crater" with a diamond hoard ten times bigger than known global reserves, enough to "supply world markets for next 3,000 years."[23] This might be good news to diamond-starved consumers, unaware that if diamonds begin to flow like water, the subjective theory of value will soon kick in. But Rhodes and Oppenheimer – if they believed the Siberia story at all – would probably find it depressing: yet another argument in favor of the cartel that so many inside and outside the industry had fought to bring down.

CHAPTER THREE

Blood Diamonds

Illicit diamonds

In the late summer of 1967, I arrived in Sierra Leone to take up a teaching position at Koidu Secondary School. Located in Kono District in the country's Eastern Province, Koidu was and still is the center of Sierra Leone's diamond country. Then, it was a Wild-West town, said to have a population of 5,000. The true number was more like 20,000, most of the difference made up of people who had come for diamonds. There were an estimated 10,000 "san-san boys" – illicit diamond diggers, so-called because they spent their days digging and sifting "sand" without a permit. In fact there may have been many more. Some estimates placed the total number of san-san boys roaming over the country's diamond fields at 100,000. Some were boys; most were men. Many were children. In Koidu there were also some 2,000 "Lebanese" residents, although "Lebanese" – interchangeable with "Syrian" – included anyone of Middle Eastern appearance, including the Indian proprietors of the Chellaram and Chanrai general stores.

Many of the Lebanese were second- and third-generation Sierra Leoneans, the progeny of an exodus from the collapsing Ottoman Empire in the early years of the twentieth century. Most of them and the san-san boys, and many more among the crowds that coursed through Koidu, were there illegally. To reside in Kono District you had to be a native Kono speaker or you needed a government pass, because large parts of the district – 230 square miles – had been leased to the Sierra Leone Selection Trust (SLST) for its exclusive use. SLST, part of the Selection Trust empire, then one of the biggest mining conglomerates in the world, sold all of its diamonds to De Beers. And SLST had Sierra Leone, and especially Kono District, sewn up. Or so, from the safety of Yengema, their company town 29 miles down the road, its representatives liked to think. Yengema was everything Koidu was not. It had paved roads, electricity for company operations, an airstrip for the shipment of diamonds, and on the company compound there was a shop where a privileged few could purchase frozen New Zealand lamb chops, French wine, and Irish butter. There was even a scruffy nine-hole golf course.

Koidu was different; it was dirty and lawless. The night I arrived, three san-san boys were killed, apparently with their own digging implements. Murder by shovel. Violence was common, theft an everyday occurrence. There was evidence that the roads may once have been paved and lined with storm drains, but these were long gone. Of running water there was little, and electricity was reserved for those on a tiny grid and the few others with private generators. SLST tried to keep a lid on Koidu and the diamond fields beyond with an army of 500 uniformed company police, a fleet of trucks to move them about, and two Bell 47G4 spotter helicopters. The police roamed the towns and villages looking for illegal diggings, rounding up san-san boys, and carting them on a regular basis 100 miles or so down the road where they were dumped in the forest. Most would be back inside a week. I knew, because students were sometimes caught up in a raid and – innocent or not – they too would be back in a few days.

I recall stepping into an open-fronted Lebanese dry goods shop one day shortly after arriving in Koidu, with the aim of buying a couple of shirts. All of the shirts were in cellophane wrappers, and all of the wrappers lay under a heavy coating of red dust kicked up by passing traffic. The shop owner was obviously not interested in selling shirts, or in dusting. He offered me some thick Turkish coffee while keeping a sharp eye on the road for genuine clients – san-san boys with something to sell. He told me, in the friendly way an uncle might speak to a dim nephew, that if I wanted shirts I should find a tailor somewhere.

He was part of the Lebanese diamond mafia – about which, more later. They operated in plain sight, buying diamonds and moving them out of the SLST lease area for smuggling abroad or for sale to slightly more legitimate buyers operating down-country.

The Lebanese needed help, of course, and there were many willing hands. Army officers had staged Sierra Leone's first military coup in 1967, one day short of the sixth anniversary of independence. A year after that, the Principal of my school attended a ceremony in Yengema on the seventh anniversary. He sat in the audience behind two army officers who were comparing diamonds they had picked up since their arrival the day before. On New Year's Day, 1968, I sat on the veranda of my house at Koidu Secondary School and listened to a growing clamor from the center of town. Someone had discovered a diamond by the side of the town's fetid lake, and a rush was in progress. Men, women, and children dug up mud and whatever sewage it contained; housewives carted head-loads of it away on kitchen trays; men shovelled ooze into the trunks and back seats of taxis and buses until their springs snapped.

Everything in Koidu was about diamonds. Twice I had students bring me diamonds before class, just as students in old movies bring the teacher an apple. The diamonds weren't much and I soon lost them, but all the students knew about two of the Four C's – color and carats. My diamonds were said to be worth 10 cents.

Henneh Shamel, scion of a prominent Shi'ite family of diamond dealers, once introduced himself to me in a bar, perhaps because I stood out (I was white – still am), and after he left there were furtive whispers about his diamond business. Sometime later, in 1968, there was a spectacular diamond robbery at Hastings Airport in Freetown. A gang of armed men met the SLST De Havilland Heron as it flew in from Yengema and they stole $3.4 million worth of diamonds. Shamel was charged with the crime. He was tried, but he was acquitted for lack of evidence, and he had the sense soon after to leave Sierra Leone as it began its long slide into anarchy, madness, and war.

I tell this story to give a flavor of the chaotic atmosphere of artisanal diamond mining and the difficulty in controlling it. Botswana, with its democracy and good governance, is often held up as a model of what Sierra Leone could have been. But Botswana's diamonds are all in kimberlite pipes that remained undiscovered and undisturbed until the 1960s. The footprint at the surface was small and easily secured. There were no alluvial diamonds scattered over hundreds of square miles and there was no way for poor people to invade the concessions of the mining company. Sierra Leone could certainly have managed itself and its diamond resources better. But even if it had been so inclined, it would have been working against a stacked deck, as were almost all of the other countries in Africa afflicted with alluvial diamonds.

From entrepreneurial to organized crime

The diamond industry was, until the beginning of the twenty-first century, almost completely unregulated. This may seem like a rather sweeping and perhaps contentious generalization. De Beers – with its virtual monopoly on the industry for most of the twentieth century – did everything it could to control mining and the trade in rough diamonds. It spent millions of dollars every year on security operations in a dozen countries, and where it could not halt the flow of smuggled goods out of a country like Sierra Leone, it set up buying operations across the borders in Liberia and Guinea in order to mop up the leaks. De Beers and its sister companies hired all manner of police and security officers to keep the business under control. Fred Kamil, an unstable and unscrupulous Lebanese soldier of fortune, was one such individual, creating a small army of mercenaries on the Sierra Leone – Liberia border in the 1950s to "dissuade" diamond smugglers. He later hijacked an aircraft in Southern Africa, hoping to force an interview with Harry Oppenheimer but instead he wound up in prison in Malawi.

One man who did get an interview, not with Harry but with his father, Sir Ernest, was Percy Sillitoe, former head of MI5. Sillitoe, who had as a young man worked for the British South Africa Police – becoming "familiar with cases of diamond smuggling" – was brought out of retirement to head up a new De Beers operation, the International Diamond Security Organisation (IDSO). Sillitoe, a heavy-duty Cold Warrior, was intrigued by Oppenheimer's tales about a communist-directed diamond-smuggling ring. The "ring" stole from De Beers, SLST, and others, and shipped diamonds from West Africa into the Soviet Union where they were being used to make H-bombs. Diamonds were also being used to finance anti-West uprisings in Greece, Lebanon, Syria, and Algeria. In 1957, Ian Fleming, with five James Bond novels already under his belt, got in on the act, writing a non-fiction book called *The Diamond Smugglers*. In it, he detailed the "fantastic but true story of the world's greatest smuggling racket." As a result of the escalating political excitement, Sillitoe was able to call on the British government for cash and the loan of MI5 agents as he pursued the communist conspiracy.

In fact, however, most of Fleming's book, while definitely fantastic, was not true. There was no communist plot, for example. By 1953, the Soviet Union had its own diamonds, and it certainly had no need of gem diamonds for "H-bombs" or any other industrial purpose. In any case, De Beers would soon make a deal with the Soviets to buy and resell *their* diamonds, so there was no need to worry about gems moving in the other direction. The creation of the IDSO had one major purpose: to curtail diamond smuggling from Sierra Leone into Liberia. The lesser strategy was to identify buyers and shippers in Monrovia and to put them out of business (or buy their goods). The more important part of the game was to frighten the British government with such a large communist fantasy that it would change its tight post-war foreign exchange rules in Sierra Leone, allowing De Beers to pay better prices in hard currency for the diamonds that were going south. Although Fleming, Sillitoe, Kamil and others wrote enthusiastic memoirs about their cloak-and-dagger exploits, only six people were ever arrested in the entire IDSO operation, which was quietly brought to an end after only three years.

In reality, a much more dangerous game was about to begin. In 1968, after two more military coups, a civilian government was reinstalled in Sierra Leone, headed by a trade union populist, Siaka Stevens. Stevens lost little time in ridding the country of his political foes and declaring a one-party state. He corrupted the army and police, he suborned the judiciary, and he forcefully suppressed all dissent. He then created the National Diamond Mining Company (NDMC), essentially nationalizing SLST. All important decisions in the NDMC were made by President Stevens and an Afro-Lebanese businessman, Jamil Sahid Mohammed. Together, they and their henchmen pushed Sierra Leone's diamond industry into the shadows where it could be systematically looted. In 1970 Sierra Leone exported more than 2 million carats. By

1980 official exports had dropped to 595,000 carats, and in 1988 only 48,000 carats left the country through legitimate channels.[1]

During the 1970s and 1980s, various factions in Lebanon's violent civil war sent fundraisers to Sierra Leone and other African countries where the Lebanese diaspora worked the diamond trade. PLO leader Yasser Arafat visited Sierra Leone looking for support, and this attracted Israeli attention. During the 1980s, a succession of Middle Eastern, Russian and American mobsters trolled through Sierra Leone where each in turn took as many diamonds and as much of the country's patrimony as they could, corrupting all in their path. By the early 1990s, Sierra Leone's economy lay in ruins and its government had become little more than a predatory menace in the lives of its people. The one constant through all of it was diamonds. Diamonds, which had never fueled much in the way of development, continued to be mined, and they now fueled what was left of an almost completely criminalized government.

From organized crime to blood diamonds

A military coup in April 1992 momentarily raised the hopes and expectations of ordinary Sierra Leoneans when its leaders promised to end corruption. But before long the new rulers too were snout-down in the diamond trough. A year before, something else had happened. A small band of guerrillas, calling themselves the Revolutionary United Front (RUF), had crossed from Liberia, attacking and looting two border villages. Speaking later on the BBC, their leader, Foday Sankoh, announced the beginning of a "people's struggle" that would end corruption and bring democracy to Sierra Leone. Sankoh, however, had neither the experience nor the temperament to carry out either of these aims. A former army corporal jailed in a coup attempt many years before, he claimed to be a photographer by profession. If he took photographs, few remain, but there is clear enough evidence of where his backing came from.

Sankoh's first source of support was Libya's Colonel Muammar Gaddafi. During the 1980s, Gaddafi was aiding and abetting all manner of revolutionary behavior in Africa, bringing students, dissidents, and almost anyone looking for trouble to his World Revolutionary Headquarters near Benghazi for "ideological training." Coup-makers from Burkina Faso and Gambia spent time there, as did Laurent Kabila, who would eventually oust Mobutu Sese Seko from the Congo. A young Liberian dissident, Charles Taylor, was there, perhaps coincident with the visit of Foday Sankoh and other Sierra Leoneans. Gaddafi would bankroll both Taylor and Sankoh in the years ahead, but both of them wanted and needed more support than Gaddafi was prepared to offer.

Taylor started his own war against the Liberian government in 1989, and by the time Sankoh was ready to move against Sierra Leone, Taylor was able to offer him a Liberian base, supplies, training, and men. At the time, Taylor was developing a new fundraising strategy based on natural resources. In fact, what Taylor started in Liberia would become the fundraising tool of choice for a coming generation of African warlords. He began with timber. Liberia had vast stands of untouched tropical rainforest, and, in a world of dwindling supply, rising prices, and growing environmental concern, Taylor saw unalloyed opportunity. In 1990 his National Patriotic Liberation Front (NPLF) captured the Port of Buchanan, using it to import weapons and export timber. When peacekeeping forces blockaded the port, he turned for assistance to the President of Côte d'Ivoire, who supplied him with more weapons and allowed him to ship timber and iron ore through the Ivorian port of San Pédro.

But the diamonds of Sierra Leone beckoned. It takes little genius to know that diamonds are smaller and lighter than timber and iron ore. In fact they have the highest value-to-weight ratio in the world. Liberia had few diamond resources of its own, but for decades it had been an outlet for illicit goods from Sierra Leone. Instead of waiting for diamonds to trickle reluctantly across the border into war-torn Liberia, however, a new possibility arose: send in proxies to bring the diamonds out. Foday Sankoh may have talked about a "people's struggle," and he certainly aimed to take power in Sierra Leone, but once his RUF gained momentum, they headed for Koidu where I once taught high school, and the diamond fields of Kono District.

By 1995 they were firmly ensconced, and although the battle for Freetown, 270 kilometers to the west, would ebb and flow in the coming years, the RUF held onto Kono and its diamonds for most of the war. And most of the diamonds they mined, either themselves or using forced labor, were exported through the offices of the warlord and – from 1997 – President of Liberia, Charles Taylor. Estimates of the value of RUF diamond exports vary between $25 million and $125 million per annum at the height of the plunder. The actual number is probably between $60 million and $80 million,[2] but even if it was a fraction of that, it would have bought a lot of weapons in a world overflowing with used AK-47s, and with gunrunners aplenty to supply them.

The RUF's stated aim to end corruption and bring democracy to Sierra Leone never rang true, and fortunately they never took power, but the damage they inflicted was catastrophic. They waged their war largely against civilians, terrorizing towns and villages by chopping the hands and feet off civilian adults and children. The tactic served as a warning to any who might oppose them, and it cleared vast areas – including the diamond fields – allowing the RUF to forage with impunity.

They perfected the art of child soldiering, enlisting girls as sex slaves and boys as killers, socialized into violence through drugs and murder, often of their own parents. Half the country's population was displaced; millions moved into camps and across borders. The death toll is estimated at 75,000, but that number is certainly low. The RUF ransacked Freetown twice and the conflict raged on, longer than World War I and World War II combined. Schools and hospitals closed; farmers abandoned their fields. Sierra Leone, already last on the UN Human Development Index, ceased to exist as a functioning country. Here was the blood diamond phenomenon in its most basic, brutal form.

Looking at resource wars, scholars have asked how much these conflicts have to do with greed, and how much with grievance. Diamonds did not cause the war in Sierra Leone, and few of the RUF gangsters who pillaged the countryside got rich. Greed was not a major factor for them, and they undoubtedly had grievances – as did the Brownshirt thugs of the 1930s and Pol Pot's Khmer Rouge in the 1970s. But Sierra Leone's conflict resists the definition of a civil war because the RUF fought against everything and everybody: the government, civilians, children, a West African peacekeeping force, and eventually the very tardy United Nations peacekeeping force that finally arrived in 1999. Fittingly, one of the RUF incursions into Freetown was called "Operation No Living Thing." The RUF had no ethnic basis; their fight was not about land and it had no focussed ideology. And unlike many conflicts of the late twentieth century, theirs had no Cold War roots. It was about power, pure and simple. It was a conflict made possible – the RUF would have said *necessary* – by the corruption and incompetence of a diamond-addled government. The irony, of course, is that the RUF would use the same thing that created the governmental venality it hated – diamonds – to wreak a staggering level of additional brutality on an already ruined country.

Angola

Conflict broke out in Angola long before it did in Sierra Leone, but, at the start, diamonds played only a minor and indirect role. Diamonds, however, as well as the country's other raw materials and agricultural produce, were part of the calculus that caused Portugal to cling to Angola and its other colonies long after other European powers had hauled down their flags. In 1960 there were almost 180,000 Portuguese settlers in Angola, and more were on their way.

Angola's first diamonds were discovered in 1912 in the northeastern province of Lunda, and later in the Cuango Valley to the west. At first the diamonds coming out of Angola were alluvial in nature, but kimberlite pipes were later discovered, the largest at Catoca, which remains the world's fourth-largest diamond mine. Unrest – the product of oppressive Portuguese policies, deep poverty, and forced labor – grew during the 1950s and burst into flame in 1961 with the start of a guerrilla war for independence. It would be 13 years before Portugal understood that it could not stay, and during that time, the malignant seeds of Angola's future misery took root.

Three guerrilla armies fought against the Portuguese, each vying for power on its own terms, each with its own Cold War patron or patrons. The MPLA (Movimento Popular de Libertação de Angola) was backed by the Soviet Union and ranged across the north of the country. The Frente Nacional de Libertação de Angola (FNLA) was supported by elements in the newly independent Congo, France, Israel, and Germany, and later China. In 1966, a young FNLA leader, Jonas Savimbi, broke away to create the União para a Independência Total de Angola (UNITA). Each of the movements had distinct ethnic roots, and each had a Marxist ideology, or at least a Marxist publicity department to indulge their principal patrons.

The Marxist rhetoric dampened whatever Western

enthusiasm there might have been for Angolan independence. But there were other factors. The Anglo-Portuguese Treaty of Windsor, dating from 1386, was (and remains) the oldest diplomatic treaty in the world. It had proven useful to Britain during World Wars I and II. And Portugal, a founding member of NATO, hosted an important US air base in the Azores. In short, Portugal's Western allies were, in a Cold War world, unlikely to offer more than faint disapproval for the escalating conflicts in Angola, Mozambique, and Portuguese Guinea.

It was not the war, but a 1974 coup in Portugal that changed the geopolitical calculation. When the new Portuguese government announced its decision to leave Angola, the three liberation movements began to rearrange their ideologies. Each knew that the first army into Luanda would be hard to dislodge; Luanda was the prize, and each needed help. The FNLA added American support to its Chinese backing, while the Soviet Union increased its support to the MPLA. Cuba entered the fray, sending the MPLA money, weapons and troops. Jonas Savimbi dropped his Marxist rhetoric and his Chinese backing, opting for a pro-West, multiracial stance that quickly drew direct military support from South Africa.

When the sun rose over Luanda on Independence Day in November 1974, MPLA forces occupied the city and the MPLA became the de facto government. The peace that ensued, however, was fleeting. Within two years, the war had reignited, but this time the factions battled against each other. The MPLA received massive infusions of cash, weapons and soldiery from its allies, and by 1985 there were an estimated 45,000 Cuban troops and 950 Soviet officers in the country. This was apartheid South Africa's worst nightmare come true: communist hordes on its very doorstep.

Jonas Savimbi offered South Africa the answer it so desperately wanted, and when Ronald Reagan moved into the White

House in January 1981, he sympathized. American support for UNITA up to then had been indirect and discreet. That now changed with vast new infusions of money, training, and supplies. Savimbi quickly became a contender, a credible military alternative to the MPLA and its communist backers. The money and weapons flowed, and Reagan called Savimbi, the one-time Marxist revolutionary, "the Abraham Lincoln of Africa." Angola was now the archetypal proxy war in the struggle between East and West. But while Cubans and South Africans fell in battle, by far the greatest number of victims were Angolan.

It is hard to know what might have happened had the Soviet Union not started to fall apart in 1990, but when the Berlin Wall came down, it was obvious to the Angolan factions that a new game plan would soon emerge. In 1991 both sides agreed to a time-out. Soviet money and Cuban troops disappeared with the collapse of the Soviet Union. Nelson Mandela had been released from prison and South Africa was pulling in its horns. And the United States was fast losing interest in Jonas Savimbi. Besides, under the MPLA, Angola was pumping almost 500,000 barrels of oil a day to an oil-thirsty world, something that helped focus international attention in other ways.

This was nothing if not bad news for Savimbi, especially after UNITA lost the 1992 elections he had expected to win. His domestic support waning and his international support gone, Savimbi turned to an asset he had dipped into off and on since the 1970s: diamonds. In 1984 UNITA took control of key diamond areas in the Cuango Valley, exporting $4 million worth that year. The following year it disrupted government sales by overrunning the state-owned diamond sorting facility at Andrada, and by 1993 it had taken over the best diamond-producing areas in the country. Through most of the 1990s UNITA was said to be exporting well over $1 million worth of diamonds a day,[3] running a "ministry" of natural resources and holding tenders for international buyers at its Andulo headquarters. The diamonds paid for weapons – not just AK-47s, but sophisticated armaments, rockets, and even tanks, flown in by gun-runners and the fly-by-night airlines that operated with impunity across much of Africa.

Angola was a humanitarian catastrophe. Half a million people were killed and a third of the country's population was displaced. Whatever rudimentary services had been left by the Portuguese were wrecked. Poverty, disease, and soaring levels of child mortality were the norm. By 2002 the warring factions had laid an estimated 15 million landmines.

As in Sierra Leone, diamonds did not cause the war in Angola. But, without them, the conflict would almost certainly have ended a decade before it did. Savimbi's access to weapons after the collapse of the Berlin Wall was almost exclusively financed by the sale of diamonds into a trading system that paid no attention and asked no questions. Before UN sanctions came to Angolan diamonds, De Beers and others were eager recipients of whatever they could lay their hands on. Through much of the 1990s, De Beers annual reports spoke enthusiastically about the company's ability to purchase better-quality Angolan diamonds openly through its outside buying offices. Once sanctions were applied, the diamonds continued to move with relative ease through the hands of greedy, sanction-busting African heads of state, gun-runners and the wide assortment of criminal bottom-feeders that infested the industry.

Congo

In the vast array of tragedies that mark Africa's long and complicated relationship with Europe, the Congo is perhaps the worst. Late to the scramble for African colonies, Leopold II,

King of the Belgians, set his sights not so much on the African coastline that had already been marked out by Britain, France, Germany, and Portugal, as on the vast, uncharted interior. In a few short years during the 1870s and 1880s, Leopold was able to lay his hands on a territory 77 times the size of Belgium, calling it the Congo Free State. Anything but free, the Congo served as Leopold's personal fiefdom until it was prised from his bloody grasp in 1908. The irony of his rule is that it began with widespread international praise for his stated aim of fighting African slavery. In reality, he instituted a kind of slavery of his own – forced labor, plunder, cruelty, and untold violence against all who resisted his will. The Congo made Leopold rich beyond imagination, but for the people he ruled, he was a catastrophe. A 1919 Belgian government inquiry found that, between the arrival of Leopold's private army – his Force Publique – and its departure, the population of the country declined by half. Something between 10 and 13 million people died of war, disease, starvation, and exhaustion. Leopold's rule was a human rights crime of genocidal proportions.

When the Belgian government took control of the Congo, matters improved, but Belgian colonial rule too focussed almost exclusively on the extractive sector and the colony's natural resources. In the late 1950s, Brussels was caught completely off-guard by the winds of change that had started blowing across Africa. When it came, independence was so sudden, and all parties were so unprepared, that calamity was an almost foregone conclusion. The colonial infrastructure, such as it was, had served Belgian enclaves and Belgian needs. Institutions of governance were veneer-thin, and after 50 years of direct Belgian rule, there were only 17 Congolese university graduates to fill the vacuum. The army mutinied within a week of independence in June 1960, and, a few days later, mineral-rich Katanga Province seceded. The UN was reluctant to grapple with secession, so the country's new Prime Minister, Patrice Lumumba, asked for Soviet assistance. In doing so, he wrote his own death warrant. Within three months he was deposed and a few weeks after that, with the direct involvement of Belgium and the United States, he was executed.

For most of the next four decades, the Congo would be ruled by the man Lumumba had made his army Chief of Staff, Joseph Desiré Mobutu. Mobutu held power through a toxic mix of plunder, violence, and Cold War guile. In an "authenticity" campaign, he renamed himself Mobutu Sese Seko and he renamed the country Zaire. He created a one-party state, jailing, executing, and bribing his rivals into submission, and he subordinated every aspect of state apparatus to his personal rule. He pitted political enemies, foreign investors, and Cold War antagonists against one another as a cat plays with mice, and, for much of his rule, diamonds were a centerpiece in the game.

Congolese diamonds, both alluvial and kimberlite, were discovered in 1907 in vast stretches of territory around Kisangani, Mbuji-Mayi and Tshikapa. Mining operations fell under the authority of the Société Internationale Forestière et Minière du Congo, known as Forminière, and by 1929 – after South Africa – Congo had become the second-largest diamond producer in the world. The diamonds, however, were largely low-quality "bort," and until the 1930s when an industrial use for them was developed, they had little value. The growing machine-tool industry, World War II, and a voracious demand for weapons changed all that. Ernest Oppenheimer wrote to his son, Harry: "Forminière will dictate the post-war politics of the diamond trade. By controlling the Congo production, De Beers will maintain its leading position in diamonds."[4] The deal was straightforward enough: Belgium would sell the entire Congolese production to De Beers and, in return,

De Beers would ensure that Antwerp was well supplied with gem-quality diamonds for its cutting and polishing industry.

De Beers maintained its monopoly buying position in the Congo into the 1970s, taking all of the production of Forminière's successor company, the Société Minière de Bakwanga (MIBA). Mobutu's plunder began with copper and other mineral resources, but in 1973 he turned his attention to diamonds, nationalizing MIBA. De Beers kept its buying monopoly until 1981, but MIBA production fell by 35 percent during the intervening years. Many of the better diamonds simply crossed the river to Brazzaville where they were re-branded as a product of the Republic of Congo.[5] In 1981 Mobutu "liberalized" the diamond sector, promoting alluvial mining and allowing MIBA to be looted by his cronies. But in forcing the domestic price of diamonds down in order to increase his margins in Antwerp, Mobutu simply cut off his nose to spite his face. The flow of better diamonds across the river to Brazzaville increased, and they went as well to Zambia, Angola, Burundi, and any other place where better prices prevailed. Diamonds nevertheless remained an important part of the Congolese economy, representing as much as one-third of the country's formal exports into the mid-1990s. And they were an essential part of Mobutu's personal economy. He gave diamonds as gifts to French President Giscard d'Estaing, and – always a friend to the West – he helped Jonas Savimbi in his war against the Marxist MPLA, in later years acting as a conduit into world markets for smuggled Angolan diamonds. An indication of the level of corruption can be seen in 1995 Congolese and Belgian diamond statistics. Officially the DRC exported $331 million worth of diamonds to Belgium that year, while Belgium imported $646 million from the DRC.[6] It is unlikely that either figure bears much resemblance to the truth.

By then, however, the endgame for Mobutu was drawing

close. The Rwandan genocide that took place in 1994 ended with the flight of an estimated 1 million Hutu men, women, and children across the border into the DRC. With them were the remnants of the armed and genocidal Hutu militia, now part of a new Mobutu calculation. The military and criminal possibilities here, however, were more than even he could manage, and in 1996 the new government of Rwanda and its ally Uganda began to provide a coalition of Congolese rebels with weapons, logistics, and cash. Angola, tired of Mobutu's support for UNITA, turned on him as well, as did other neighbors.

The Alliance of Democratic Forces for the Liberation of Congo-Zaire (AFDL) was not much of an alliance and it was anything but democratic. But in its aim to liberate the Congo from Mobutu, success came quickly. In 1997, under the leadership of long-time rebel leader Laurent-Desiré Kabila, the AFDL moved swiftly, taking the diamond towns first: Kisangani, Mbuji-Mayi, and then Lubumbashi. To finance his march on Kinshasa and no doubt much more, Kabila sold mining concessions to all comers, and in May 1997, four days after Mobutu fled aboard a plane provided by Jonas Savimbi, he entered Kinshasa.

As soon as Kabila was in the driver's seat he began to look in the rear-view mirror, regretting some of his hasty salesmanship – notably where diamonds were concerned. Taking a leaf or two from Mobutu's driving manual, he banned foreigners from the diamond trade and instituted a new set of taxes and fees. All traders were required to pay a $3 million membership fee in a newly created Kinshasa diamond bourse, and all transactions had to take place in (worthless) Congolese currency. Not surprisingly, diamond sales collapsed. Or rather legitimate diamond sales – which had picked up after the departure of Mobutu Sese Seko – collapsed. In response, and to make matters worse, Kabila changed the rules again, giving a monopoly on diamond exports to an Israeli firm, International Diamond Industries (IDI). How much money changed hands is not known, but IDI, understanding that it might have only months, if not weeks, to make a return on its investment, drove buying prices down. A real monopoly – De Beers, for example – might have been able to make this work. But IDI did not control the police, or diamond miners, traders, or even Laurent Kabila. And it certainly did not control traffic on the River Congo.

Congolese diamond exports dropped from $20 million a month to either $3.7 million or zero, depending on which Congolese statistical department you believe. Belgian imports fell from $50 million in September 2000 to $24.6 million two months later in November. Neither the Congolese nor the Belgian numbers are much more than fanciful. Both countries suffered from a surfeit of smuggling, false invoicing, transfer pricing, and other devious behavior, but they do show that the IDI monopoly was largely worthless, accomplishing little more than a diversion of Congolese diamonds into other channels. The Republic of Congo, which lies to the northwest of the DRC (and is sometimes distinguished as "Congo-Brazzaville"), and which has few diamond resources of its own, had long been a favored route for the avoidance of Mobutu and his kleptocracy, and that simply continued. Like Liberia to the north, Brazzaville had become a major diamond laundering center, accounting for as much as 10 percent of world production in 1996.

Meanwhile, Kabila seemed hell-bent on making as many enemies as he possibly could. He annoyed Rwanda and Uganda, his erstwhile backers, by incorporating Hutu militia into his army. Soon Rwanda and Uganda were supporting anti-Kabila rebel armies. Zimbabwe and Angola came to Kabila's rescue, and within months a proxy war was being fought among a collection of armies and rebels, each with territory and mineral resources in mind. Rwanda and Uganda focussed their attention on Kisangani, and frequently clashed in their efforts to loot diamonds. An estimated $70 million worth of diamonds left Kisangani in 1999 under the auspices of foreign and rebel armies. Belgian customs duly noted the import of millions of dollars' worth of diamonds from Uganda and Rwanda, countries with no diamonds of their own. The Rassemblement Congolais pour la Démocratie, a prominent anti-Kabila rebel movement led by Jean-Pierre Bemba, used whatever came to hand to finance its war. Gold was a favorite, as was columbite-tantalite – coltan – used to make capacitors in cell phones, computers, and almost every other kind of electronic device. Much of the contraband went out through the Rwandan capital, Kigali. Over one 18-month period, Rwanda exported $250 million worth of coltan, a mineral only available at the time in the Congo. Bemba, like so many others, was also fond of diamonds, recording shipments worth $4 million in 2000. The numbers, of course, represent a fraction of what was actually going on.

Rebels and their allies were not the only predators. Countries that went to Congo in order to help Kabila stayed on to help themselves. Zimbabwean military officers set up a company called Operation Sovereign Legitimacy (OSLEG), aimed at exploiting diamonds. OSLEG eventually went into business with a Congolese parastatal and a company called Oryx, based in the Cayman Islands. When they were accused of smuggling, money laundering and outright theft by a UN Panel of Experts, all parties oozed righteous indignation. Oryx denied the charges, saying that it had a "strong commitment to helping local communities by building schools, roads and bridges. It provides fresh water, donates food and anti-malaria drugs, and is working with the World Health Organization to develop a tsetse fly eradication program."[7] Zimbabwe's Defence Minister, Moven Mohachi – later killed under suspicious circumstances – said, "We saw this as a noble option. Instead of our army in Congo burdening the treasury for more resources, which are not available, it embarks on viable projects for the sake of generating the necessary income."[8]

Viable projects, yes; noble, not so much. A lot of people became very wealthy, but few were among the ordinary rank and file of Congolese citizens. Others who came to do good and wound up doing pretty well for themselves were members of the Namibian armed forces, who used a Namibian parastatal called August 26 Holding, created by the Ministry of Defence, to exploit diamonds in Tshikapa. And the Central African Republic (CAR) became involved as well, acting as a conduit for supplies to Jean-Pierre Bemba's newest rebel coalition, the Mouvement de libération du Congo (MLC). Bemba paid in whatever kind of currency he could dig up. Diamonds, of course, had become the hard currency of choice for rebel movements, and a comparison of Belgian and CAR numbers tell the tale. In 2000, the Central African Republic recorded official diamond exports of 461,000 carats, while Belgian customs recorded inbound CAR diamonds totalling 1.3 million carats. The difference represented about $90 million per annum between 1996 and 2001.

The venality, corruption, and violence around the Congolese diamond trade was stunning. Like frogs in water coming slowly to a boil, industry leaders seemed either not to notice, or not to care. When stories like these started to become public, a common refrain in the diamond industry was: "Diamonds don't kill people; guns kill people." This is not unlike the National Rifle Association's slogan, "Guns don't kill people; people kill people." It's tempting to make a joke out of such foolishness, but the subject is too serious. Using some of the world's leading epidemiologists, the International Rescue Committee conducted a series of detailed studies not unlike the Belgian investigation of 1919. They concluded in 2007 that an estimated 5.4 million *excess* Congolese people had died from conflict-related causes since 1998. Mostly these were not combat deaths. They were deaths from preventable and treatable disease: malaria, diarrhea, pneumonia, and malnutrition.[9] And of violence, of course, there was no shortage, much of it aimed at civilians, and much of it at women. Rape became a common terror tactic, not unlike the RUF's chopping off of hands and feet in Sierra Leone. A 2010 study by representatives of the International Food Policy Institute, Stony Brook University, and the World Bank found that as many as 1.8 million Congolese women had been raped during their lifetime, and over 400,000 in the previous 12 months.[10]

Al Qaeda

By the mid-1990s, diamonds offered such a risk-free method of laundering money, you would not have been much of a terrorist organization if you hadn't given them some thought. That seems to have been what al Qaeda did. It might have come to the idea through the Lebanese diaspora and fundraising efforts undertaken among African diamond dealers by factions in the Lebanese civil war. Or al Qaeda operatives might have rubbed shoulders with students at Colonel Gaddafi's World Revolutionary Headquarters in Benghazi. One of these was Ibrahim Bah, a senior RUF leader in Sierra Leone who had in earlier years fought with Hezbollah in Lebanon's Bekaa Valley, and later with the Mujahidin against the Russians in Afghanistan. Bah was a bagman who organized and managed the trade of diamonds for weapons through Liberia. In a 2001 *Herald Tribune* article, the paper's West Africa correspondent, Douglas Farah, exposed a three-year diamond-buying relationship between al Qaeda operatives and the RUF, working through Liberia. Farah identified three men: Ahmed Khalfan Ghailani and Fazul Abdullah Mohammed spent time in

Liberia and Kono District in Sierra Leone in an arrangement set up in 1998 by a third man, Egyptian national Abdullah Ahmed Abdullah. All three were linked to the 1997 bombing of US embassies in Kenya and Tanzania, all three were members of al Qaeda, and all three were on the FBI's list of Most Wanted terrorists.

Farah's story, based on unnamed European intelligence sources, broke only three weeks after 9/11, and the RUF quickly bent over backward to deny it and any other al Qaeda connection. The denials smacked of protesting too much, but it is possible, even probable, that men like the three in question would not have shown the RUF their al Qaeda ID cards. Charles Taylor also vehemently denied any al Qaeda connection, and, by way of showing his displeasure, so threatened Farah that the *Herald Tribune* pulled the reporter out of Africa for his own safety. Clearly, nobody in West Africa wanted to be caught in the 9/11 mangle. In due course, the 9/11 Commission, which could have delved into the issue, did not, and the names of these three men do not appear in the final Commission report. Mike Shanklin, however, who headed CIA operations in Liberia during the 1990s, said "Al Qaeda, Bah, Taylor, they were there. There is no question in my mind that those people were there. They were there during the period in question. And clearly they were involved in some sort of diamond business. That's a fact."[11]

Ghailani was captured by Pakistani forces in 2004 and sent to Guantánamo Bay in 2009. He was tried in a New York City court for his role in the embassy bombings, and, during the trial, tales of his visits to Liberia surfaced again. He was convicted of conspiracy in 2010, sentenced to life imprisonment, and now resides at the ADX "supermax" prison in Florence, Colorado. Fazul Abdullah Mohammed was killed in a 2011 firefight when the car in which he was traveling near Mogadishu was stopped at a Somali military checkpoint. US

Secretary of State Hillary Clinton described the killing as a "significant blow to al Qaeda, its extremist allies, and its operations in East Africa."[12] Abdullah Ahmed Abdullah remains on the FBI Most Wanted list with a $5 million reward for information leading to his arrest or conviction.

Diamonds are Dangerous. That is the title of a 1960 memoir by J. H. du Plessis, a man who served in the South West African Police diamond detective department. Diamonds may be forever, a girl's best friend, and many other agreeable things, but – as the people of Sierra Leone, Angola, and the Congo came to know during the 1990s – they can be very dangerous indeed. Anything of value, of course, will attract thieves, but a confluence of factors made diamonds especially dangerous, and especially in Africa. The first factor was the alluvial nature of diamonds in Sierra Leone, Angola, and the Congo; alluvial and therefore easy to mine. And very difficult to police, especially in countries where governance had collapsed. The second was the end of the Cold War and the need among rebel armies for alternative sources of funding. Other booty – timber, gold, tantalum, tin, tungsten – has been used as an alternative to cash for the purchase of weapons. But nothing has the value-to-weight ratio of diamonds. None of the alternatives can be carried in a pocket past customs agents. None of them can be sold as quickly or as easily as diamonds.

And there was something else about diamonds. Although the industry represented several billions of dollars' worth of enterprise at the mining end of the pipeline and $70–$80 billion at the retail end, diamonds were almost completely unregulated. Belgium kept fairly close tabs on what entered the country, but only for tax purposes. It never asked whether diamonds coming from countries that didn't mine them might in some way be suspect. It never asked about the huge discrepancies between export and import values. De Beers wanted to keep its corner on the market, so if diamonds were being smuggled out of a country where it had a monopoly, it would simply post men on the other side of the border to mop up the leakage. Most companies and most governments asked no questions as the world's output was swept together for processing and then dispatched to the glittering showrooms of Paris, London, and New York.

And because there was little regulation and no independent oversight, large parts of the diamond industry became criminalized. African warlords did not have to look far to find buyers for their goods. Southbound gunrunners did not have to look far for buyers when they went home to Eastern Europe with diamonds. There was also a reverse smuggling operation moving diamonds out of Russia, laundering them through African capitals the way African warlords did. "Submarining" it was called. By the mid-1990s, at least 25 percent of the world's rough diamonds were in some way connected with money laundering, tax evasion, illicit behavior, and war. Given the fact that Angola accounted for almost a quarter of the world's supply, the percentage is probably higher. And until the blood diamond campaign began, nobody in the industry – the place where diamonds were truly understood – said a word.

CHAPTER FOUR

Activism

The United Nations

The record of UN peacekeeping efforts in Angola during most of the 1990s makes dismal reading. The effort was irregular, underfunded, and completely inadequate to the task at hand. Given the towering ambition and relentless drive of UNITA leader Jonas Savimbi, however, it is obvious in retrospect that no amount of traditional UN peacekeeping would ever have sufficed.

There were, in total, four UN missions in Angola. The first, the United Nations Angola Verification Mission (UNAVEM) was established in 1989 to oversee the withdrawal of Cuban troops. UNAVEM II, which ran from 1991 to 1995 and never had more than a few hundred observers, aimed to "verify" peace arrangements that had been agreed between UNITA and the government of Angola. Following the 1992 elections, the long-running war resumed, but UNAVEM II soldiered on in its semi-fictitious mandate. UNAVEM III was created by the Security Council in 1995 to monitor a hopeful but ultimately doomed peace protocol signed in Lusaka by UNITA and the Angolan government. This time 7,000 military peacekeepers were sanctioned, although the number never rose that high.

The Security Council issued a series of resolutions urging UNITA to abide by the terms of the peace agreement. Such resolutions typically "call upon" parties to do various things, but where UNITA was concerned, it was like calling upon the moon to vacate the sky. The war continued, especially in the diamond areas where UNITA had no intention of ceding an inch of authority or territory. Clearly the UN approach was not working, but the Security Council tried one more time in 1997 with a smaller effort, this time called the UN Observer Mission in Angola (MONUA). Its mandate was "to assist the Angolan parties in consolidating peace and national reconciliation, enhancing confidence-building and creating an environment conducive to long-term stability, democratic development and rehabilitation of the country."[1] There was, however, no peace to consolidate, no confidence to enhance, no environment in which anything but violence could thrive.

To put it in perspective, MONUA had fewer personnel than the Phoenix Police Department has cops. By the end of 1998 MONUA's numbers had dwindled to fewer than 600, and by February the following year it was gone, an admission after ten years and three-quarters of a billion dollars that resolutions, promises, observers, and standard peacekeeping methods were wholly inadequate to the task.

Among the Security Council resolutions, however, was one that acknowledged the role of diamonds. Resolution 1173 of June 1998 prohibited the purchase of any diamonds not controlled by the Angolan government. Here at last was recognition that in a post-Cold War world, one had to look beyond politics in order to understand the financing of weapons and matériel. But a UN resolution on diamonds plus $75 would get you a pretty serviceable used AK-47 assault rifle in those days, and there was no shortage of suppliers or planes to fly them. And there was no shortage of buyers for diamonds. Where Angola was concerned, sanctions were almost universally ignored, and with complete impunity.

Civil society

It would take almost two years for that to begin to change. The first element in bringing about change was an investigation of the Angolan diamond economy by the British NGO Global Witness. Part of a growing breed of campaigning NGOs, Global Witness was only five years old when it began to tackle the diamond issue. Its first investigation focussed on the illegal smuggling of Cambodian timber into Thailand; Angola was the focus of the second. In December 1998, Global Witness issued a hard-hitting report: *A Rough Trade: The Role of Companies and Governments in the Angolan Conflict*. It estimated that half a million Angolans had died in the war to that point, and that UNITA had generated $3.7 billion in diamond sales between 1992 and 1998 to pay for it. The report described the toothlessness of UN sanctions and chastised De Beers for its indiscriminate purchase of rough diamonds. The report quoted from De Beers' 1996 Annual Report, in which Chairman Nicky Oppenheimer had written about "the increasing outflow of Angolan diamonds to the major cutting centers, much of which De Beers was able to purchase through its outside buying offices."[2]

The Global Witness report explained how, with the advent of the UN embargo, diamonds were simply moved across Angolan borders into Congo, Zambia, South Africa, and elsewhere, easily disguised with a new identity and soon headed towards Antwerp, Tel Aviv, and elsewhere. "It is imperative," the report concluded, "that De Beers . . . as a major force in the world diamond trade, and the member states of the UN act decisively to end once and for all the trade that is the life blood of a guerrilla organization. If they fail to do this, they are standing in the way of the peace that the people and country of Angola so desperately need."[3]

The report might not have drawn blood without a second

important initiative that began a few weeks after it was published. In January 1999, Canada took up one of the ten non-permanent seats on the UN Security Council and it actively sought the chairmanship of the Angolan Sanctions Committee. It did this in a desire "to promote and restore the credibility of the Council's authority for the preservation of human security and the peaceful resolution of conflicts," and secondly "to curb a persistent source of conflict in Africa."[4] Canada's UN Ambassador, Robert Fowler, traveled extensively during the first half of 1999, consulting with governments and industry officials in an effort to understand how Angolan sanctions were being circumvented. He then organized the creation of a first-ever "Independent Panel of Experts" under Security Council auspices to dig more deeply into the matter. The panel traveled widely, visiting some 30 countries and speaking not only with government and industry officials, but with UNITA defectors, who provided a wealth of deeply incriminating evidence on sanctions-busting by UN member states.

Fowler listened to Global Witness and another investigative NGO, Human Rights Watch, and his final report also benefited from work done by Partnership Africa Canada (PAC). While Global Witness and Fowler focussed on Angola, PAC had been studying Sierra Leone. PAC's discovery of the diamond connection was almost accidental in the sense that its preliminary concern was the *impact* of the conflagration in that country rather than its cause. I was part of a small group of individuals who came together under PAC's auspices to discuss the war and – at first – a possible humanitarian response. At the time, I was working as an independent development consultant, while others in the group were students or individuals working for governments and NGOs. Most of those in the group, like me, had some sort of Sierra Leone connection, or hailed originally from West Africa. All were concerned that the escalating conflict was being ignored by the media, the humanitarian establishment, and the United Nations.

In one of our discussions a young Sierra Leonean-Canadian said, almost in passing, "It's about the diamonds, of course. Until someone does something about the diamonds, the war will never end." It was a Eureka moment for me, one of those self-evident observations that is not at all self-evident until someone has said it. It was a short step from there to the idea of an investigative study. The PAC study team consisted of me, Ralph Hazleton, and Lansana Gberie. Hazleton had recently retired after a long career working in some of the world's toughest conflicts, and Gberie was a Sierra Leonean journalist working on a graduate degree at the University of Toronto. As we got into the subject, we met those who had written the Global Witness report and learned that much of what they had said about Angola applied as well to Sierra Leone.

We were not rocket scientists, but we didn't need to be. It was clear enough that the RUF was holding onto Kono District, where I once taught, because of the diamonds. The *New York Times* ran an editorial that summer, spelling out what most people in Sierra Leone already knew: "Loot, not better government, has motivated the psychotically brutal guerrillas of Sierra Leone. They trade the diamonds they control for arms through neighboring Liberia, under sponsorship of President Charles Taylor, their long-time patron."[5]

Our report built on the work done by Global Witness but it differed in two ways. First, it went into more depth on De Beers, how it was structured, and how diamonds moved through the pipeline. Second, it took the lid off a dirty secret that Belgium had been hiding in plain view. For years, Belgian authorities and the industry's main body, the Diamond High Council (known by its Flemish acronym, HRD), had been assiduously recording all rough diamond imports into the country, noting whatever details an importer cared to state.

The information was freely available from open sources to anyone with an interest in the subject, and it was more than a little revealing.

We discovered, for example, that between 1994 and 1998, Belgium had imported 34.6 million carats in diamonds from Liberia, while the best guesstimates of Liberian export figures during that time – a period of almost complete anarchy and state collapse – was about 700,000 carats, 2 percent of the Belgian import figure. Massive discrepancies – not as bad as this one, but massive nevertheless – were evident in the figures for Sierra Leone, Guinea, and Côte d'Ivoire. And these discrepancies represented differences in value worth billions of dollars. Diamonds were also pouring in from countries that mined no diamonds whatsoever, among them Gambia, Burkina Faso, and Zambia.

Nobody in the HRD or anywhere else in the industry could defend the figures. There was a lame attempt at explanation in the case of Liberia: of course that volume could not possibly have been mined there. *Everyone* knew that. But we were confusing the *origin* of diamonds with their *provenance*. What we had failed to understand, they said, was that the "Liberian" diamonds were recorded as Liberian by provenance rather than origin. In other words, they had originated somewhere else and had simply transited Liberia. That line was credible for something under a nanosecond. Nobody in their right mind would have taken a cheap wristwatch into Liberia during those years, much less 34.6 million carats in diamonds.

Other campaigners were also getting in on the act. Two American congressmen, a Democrat – Tony Hall – and a Republican – Frank Wolfe – co-sponsored a bill called the Consumer Access to a Responsible Accounting for Trade (CARAT) Bill which would require forgery-proof certificates of origin for any diamond worth more than $100 entering the United States. They were backed by a group of American

NGOs, including Physicians for Human Rights, World Vision, and Oxfam America. And in Europe another coalition called Fatal Transactions was forming.

By the end of 1999, the diamond industry had started to notice the growing call for action. De Beers announced that it had placed an embargo on Angolan diamonds and it closed down all of its purchases from the outside market. But many industry players simply hunkered down and remained silent, hoping the storm would pass. Others denied the facts or denied responsibility. Martin Rapaport, an outspoken industry iconoclast, was one of the first to comment in print. Writing about Angola in November 1999, he asked,

> Is the diamond industry being used as a scapegoat, as an easy target for manipulation by the political establishment because we sell a high profile product that they think can easily be damaged by negative publicity? Should we allow ourselves to be blackmailed to support the FAA (Angolan Armed Forces) against UNITA? Are the motivations of the politicians humanitarian? Are they economic? Do they have anything to do with Angola's huge oil reserves?
>
> . . . Frankly, the whole situation stinks. I don't know the answers to these questions. I don't know if there are any answers, if there is any way to stop the wars. What I do know is given the historic role of foreign governments in Africa, it is a good idea for the diamond industry to tread with great caution and to suspect the intentions of all parties involved in this issue. We must be very careful about how we allow ourselves to be manipulated. As the saying goes, the road to hell is paved with good intentions.[6]

The PAC report, *The Heart of the Matter: Sierra Leone, Diamonds and Human Security*, was released in January 2000, causing something of a media stir in Canada, Britain, Belgium, and South Africa. In Sierra Leone, it made headlines and was the subject of a phone-in radio show. This might not seem surprising but for the fact that it was the first time there had been a public discussion about diamonds in Sierra Leone – not just in relation to the war, but ever.

When Robert Fowler tabled his report at the UN Security Council three months later, it confirmed everything that Global Witness, Partnership Africa Canada, the other NGOs, Tony Hall, and Frank Wolfe had said. And it went farther, naming for the first time in a UN report sitting heads of government – Burkina Faso's Blaise Campaoré and Togo's Gnassingbé Eyadema – as being directly and personally involved in sanctions-busting.

In the PAC report, we recommended a UN ban on Liberian diamonds. We also recommended the creation of a permanent independent International Diamond Standards Commission to establish, monitor, and enforce a code of conduct on government and corporate behavior in the diamond industry. The idea wasn't especially original. British Foreign Minister Robin Cook had already said: "The places you can sell uncut diamonds are pretty limited. It should not be beyond our wit to devise an international regime in cooperation with the diamond trade that cuts off the flow of diamonds from those who use them to buy arms and fuel conflicts."[7] And Fowler too recommended, ". . . a conference of experts . . . for the purpose of determining a system of controls that would allow for increased transparency and accountability in the control of diamonds from the source or origin to the bourses . . . including the establishment of a comprehensive database on diamond characteristics and trends."[8]

Kimberley

By now, the media were calling them "blood diamonds" and reporters were all over the story. In the PAC report we had called for an international agreement on the regulation of diamonds, but, failing that, we offered the prospect of a consumer campaign. We suggested a few possible themes:

- Diamonds are *not* a girl's best friend – witness the brutalized little girl (pictured on the cover of the PAC report) with no hands;
- "The millennium gift she'll never forget" (this was a time of extensive millennial hoopla);
- For some people, diamonds are more "forever" than for others – witness 75,000 violent deaths in Sierra Leone;
- "Diamonds are a guerrilla's best friend" – witness Sierra Leone's coups, rampaging criminals, etc., etc.[9]

Nicky Oppenheimer spoke out about the costs of a boycott to conflict-free diamond-producing countries in the developing world: Botswana, Namibia, South Africa. Nelson Mandela said the same thing: "We would be concerned that an international campaign . . . does not damage this vital industry. Rather than boycotts . . . it is preferable that through our own initiatives the industry takes a progressive stance on human rights issues."[10] In fact, however, no NGO involved in the campaign ever used the word "boycott." That word came almost exclusively from an industry that feared, quite correctly, what might happen if there was no change, and what the imagery of "blood diamonds" could do to a product sold on the basis of love, beauty, and fidelity.

Trying to understand the issue better, Martin Rapaport went to Sierra Leone in March 2000, and when he returned he wrote a powerful and unequivocal article he called "Guilt Trip."

> I don't know how to tell this story. There are no words to describe what I have seen in Sierra Leone. My mind tells me to block out the really bad stuff, to deny the impossible reality. But the images of the amputee camp haunt me and the voices of the victims cry out.

> "Tell them what has happened to us," say the survivors. "Show them what the diamonds have done to us."
>
> . . . Friends, members of the diamond trade. Please, stop and think for a minute. Read my words. Perhaps what is happening in Sierra Leone is our problem. Perhaps it is our business.[11]

The combination of NGO reports, the beginnings of a campaign, the CARAT Bill in the United States, the firestorm of negative media reports, and the industry's own growing awareness of responsibility set the stage for what came next. It began with a trip by South Africa's Minister of Minerals and Energy, Phumzile Mlambo-Ngcuka, to Britain and Canada. Mlambo-Ngcuka had been an NGO activist herself during the apartheid era, and she had an idea for a meeting that might bring industry, governments, and NGOs to the table to discuss the problem. In Britain she spoke with Global Witness, and in Canada she spoke with Partnership Africa Canada to see whether they would participate in such an exercise. She perhaps also wanted to determine whether the two NGOs were what some in the industry were now making them out to be – rabid, anti-capitalist attack dogs, inured to negativity and resistant to common sense.

We must have made a good enough impression, because, on her return to South Africa, she sent out invitations to a meeting, to be held in May, to seek solutions to the problem of what the industry and peaceful diamond-producing countries preferred to call "conflict diamonds." The meeting would bring together industry, interested governments, and civil society, and it would meet in the town of Kimberley where so much of South Africa's mineral history had been written. It was quite a meeting. NGOs rubbed shoulders with Nicky Oppenheimer, Martin Rapaport, the Belgian Foreign Minister, and delegations from a dozen diamond-producing countries. Everyone was on their best behavior. Belgium, which had denied all accusations of inadequate diamond controls, announced tough new regulations and De Beers proposed a nine-point plan aimed at keeping conflict diamonds out of the legitimate trade. There was enough congruence that Phumzile Mlambo-Ngcuka declared victory and proposed a "working group" meeting in Angola to sort out the details, to be followed two months later by a wider ministerial meeting to wrap things up.

Complexity

The Angola meeting, however, was shambolic. More governments appeared and some had second thoughts. Others arrived without instruction. Ted Sorensen, one-time speechwriter for John F. Kennedy ("Ask not what your country . . ."), showed up, claiming to represent the industry and demanding to know why NGOs were present during the discussions. Charmian Gooch from Global Witness said that the NGOs represented thousands of people and she chastised him – a man of his stature – for accepting a hatchet job from an industry so mired in excrement. After that, nobody challenged NGO participation and NGOs went on to play a key role throughout the negotiations. Despite the recriminations and indecision, the Angola meeting actually succeeded in outlining most of the elements that would be contained in the final agreement 29 months later. But it would take 11 more official meetings and many more unofficial meetings, in what soon became known as the "Kimberley Process."

The basic idea, enunciated from the outset, was that the system had to be simple, inexpensive, and based as much as possible on the existing laws of participating countries. It would take a "wholesale" approach. Instead of trying to identify, certify, and tag every single diamond, it would be based on the international shipment of "parcels," the term for packets, boxes, bags, and other containers of diamonds. Participating governments would certify that the diamonds being exported were clean, and importing countries would disallow diamonds that arrived without a certificate.

The devil, of course, dwells in detail, and details bedeviled the discussions. How should "conflict diamonds" be defined? Should there be a database of trade and production figures? Which countries should be allowed to participate? Which should be excluded? How would decisions be made? Should the system be compulsory or voluntary, and how would rules be enforced? What exactly might compliance mean? How would non-compliance be defined, and what would be done when it was encountered? Would the system have a monitoring mechanism? What role would NGOs and industry – the latter now represented by a single body called the World Diamond Council – play once the system was up and running? How would it be managed? Would there be a secretariat? How much would the whole thing cost, and who would pay?

The simple ideas proposed at the first Kimberley meeting became more and more complex, not least because some governments doubted the efficacy of the whole thing. In the film *Blood Diamond*, there is a scene representing the first meeting in Kimberley. In it the United States representative is portrayed as the leader, championing the cause. In fact, with the election of George W. Bush in 2000, the United States became one of the major impediments to an early agreement. The Bush administration was pulling the United States out of an array of multilateral agreements and there was no appetite for a new one. The real leader throughout the 29-month negotiating period was South Africa, with key supportive roles played primarily by Britain and Canada. Russia hosted one of the negotiating meetings, treating participants to a vodka-soaked dinner cruise on the Moscow River and a fashion show where models, guarded by men with assault rifles, wore mainly diamonds. I recall one of the Russians telling me proudly that the scantily clad lead model was the First Vice-Miss Russia. I guess there were several runner-up Vice-Miss Russias. Russia's major contribution to the debate was that it didn't want any statistical database in the Kimberley Process (KP). Diamond statistics in Russia were a state secret and it would take almost four years to find out why.

The NGO "coalition" was very informal. Some organizations participated at a distance; it was expensive to travel to the long string of KP meetings, and usually fewer than half a dozen NGO representatives attended. The largest numbers were European and American, although African NGOs also played an important role. PAC and Global Witness tended to lead because each devoted two or three full-time individuals to the effort and they were usually, therefore, the most knowledgeable.

The private sector took a different approach. The diamond industry is as broad as the definition of downtown, stretching from small exploration firms and giant mining houses at one end, through traders, cutters and polishers, to jewelry manufacturers and retailers ranging in size from Tiffany and Cartier to small family outlets. They came together to form the World Diamond Council (WDC) in 2000, its primary purpose to serve as the unified voice of industry as negotiations went forward. WDC leadership remained unchanged for the next 13 years, giving it the kind of memory, contacts, and continuity that most government delegations and NGOs did not enjoy.

Government participation was, in fact, patchy. In some delegations – those of Russia, Namibia, Botswana, and a few others – there was continuity. But other delegations changed with each meeting, which meant that a re-education process was needed at almost every gathering. For some governments the issue was about commerce and they were represented by their ministry of trade. For many it was about foreign affairs, while for others it was all about minerals and mining. The Swiss sent customs officials. The result was a mishmash of aims and objectives, and towards the end of the negotiations, NGOs and industry often found themselves on the same page trying to persuade a recalcitrant government newcomer of an approach they had both long since agreed.

One particularly difficult negotiating session in Brussels ended with a two-hour debate on the wording of the final communiqué. The South African Chair wanted the communiqué to state that there had been "significant progress." A point of contention during the meeting had arisen over the proposed Kimberley Certificate that would accompany parcels of diamonds. Should it be on A4 paper or 8½ in. × 11? Should it be printed in portrait or landscape format? We hadn't been able to agree on even that much, and when I said that there had been no progress at all, significant or otherwise, tears rolled down the Chairman's cheeks. I felt terrible, but I wasn't alone in rejecting the wording. Some governments objected to "significant progress" because they had come to the meeting with instructions to make sure there was no progress at all, and they didn't want a communiqué suggesting they had failed.

The issue of membership in the final scheme arose at a meeting in Botswana. It wasn't lost on many participants that a number of governments with dubious backgrounds were participating in the discussions. Togo and Burkina Faso, which had no diamond industry of their own and whose presidents had been caught in diamond-related sanctions-busting, were regular attendees. NGOs argued that there had to be a credentials committee and clear membership criteria. China agreed strongly; the United States disagreed just as vehemently, saying that the scheme should be open to all countries willing and able to join. That debate began after lunch and at midnight we were still at it. During coffee breaks, it became clear that the unspoken issue was Taiwan. Taiwan has a cutting and polishing industry and could not logically be excluded from the scheme, but China, as fixated on Taiwan as Taiwan was on China, was determined that it should not participate.

As the negotiations wore on, the NGOs ramped up the pressure. Major stories about conflict diamonds appeared in news-magazines in every language. The *Financial Times* ran regular news items based on a growing number of NGO reports and press releases. *Vanity Fair*, homeland of diamond advertising, published a story by Sebastian Junger on the war in Sierra Leone, much of it based on the original PAC report. The NGO coalition grew to more than 200 members around the world. Some followed the details carefully, and while only a handful sent delegates to the meetings, all signed petitions and translated or interpreted the issues for their own home media. In Britain, Action Aid staged an act of guerrilla theatre on the street outside a diamond industry meeting, using three actors in top hats and tails and another dressed as Marilyn Monroe in *Gentlemen Prefer Blondes*. They demanded a tough Kimberley agreement and received a huge amount of publicity. In the United States, World Vision ran a 15-second promo at the end of the season finale of *The West Wing*. It featured the voice-over of actor Martin Sheen who played President Jed Bartlett on the popular TV show. The film showed a Sierra Leonean child without hands while Sheen asked if viewers knew that diamonds were contributing to this kind of horror.

During the summer of 2000, I was drafted by the UN Security Council to take part in a new Independent Expert Panel like the one Robert Fowler had convened for Angola. This one would focus on the link between diamonds and weapons in Sierra Leone, and it included a senior Interpol officer, an arms-trafficking expert, and a man with extensive experience of air traffic control in West Africa. Diamonds were my assignment. We traveled widely, together, in pairs, and alone, to Eastern Europe, Israel, Antwerp, the Gulf States,

Southern Africa, and of course, Sierra Leone, Guinea, and Liberia.

I learned more about the origin/provenance issue during an incident in the *freilager* – the free trade zone – at Geneva Airport. In the presence of Swiss customs officials, I watched a young woman take possession of a parcel of diamonds shipped from Africa to the company she represented. She took the parcel (and us) to a tiny windowless room her company rented in the *freilager*, specially outfitted with lights, loupes, and diamond scales. She opened the package, examined the diamonds, repacked it all, and shipped it off to a new consignee in London. It was all perfectly legal, but in the space of a few moments African diamonds had been transformed into Swiss diamonds for British customs purposes, their origin permanently hidden. Once they had been sorted in Britain, they might return to Switzerland as British diamonds, and then be transhipped again to Israel or India or China – Swiss diamonds once more. Because diamonds like these never left the *freilager*, they did not technically enter Switzerland, so they were of no interest to Swiss authorities and were thus never recorded.

The Swiss maneuver was used as a means of reducing the VAT on African diamonds entering Europe. Legal, yes, but it also offered a simple and inexpensive way to obscure the diamonds' origin – a beautiful laundrette. Similar transshipments through Zambia, Gambia, the Emirates, and elsewhere made the world's customs statistics on the origin of diamonds worthless. The process offered money launderers a cheap, one-stop dry-cleaner, and the maneuver was an ideal way to hide blood diamonds.

Knowing this and trying to understand the enormous volume of diamond shipments into Antwerp from Liberia, I confronted one of Belgium's largest importers of "Liberian" goods. While some of the diamonds were almost certainly of

Sierra Leonean origin, the volumes in question exceeded even that country's production capacity. Sylvain Goldberg's hands trembled as I asked him how he explained hundreds of millions of dollars' worth of diamonds from a war-torn country that at the best of times had never produced a fraction of what he was buying. He rambled and talked of other things, and he had a large tray of rough diamonds brought in, perhaps thinking the glitter would distract me.

"The diamonds you're asking about are not conflict diamonds," he said several times; "They are not what you think. They are something else."

I never did get a straight answer from him, but a check of the Liberian street addresses on his import invoices revealed little more than cheap nameplates on dirty walls, shell operations for managing the paperwork. Unlike the *freilager* operation, the "Liberian" diamonds never even entered Liberia.

Later events would shed light on the puzzle. Goldberg and his company, Omega Diamonds, had a major share in Angola's state-owned diamond monopoly, the Angola Selling Corporation (AsCorp). Goldberg's own company was owned, in part, by a relative of Angolan President Dos Santos, and somewhere in the background lurked Arkady Gaydamak, a controversial Russian-Israeli tycoon and convicted arms dealer. In 2008, several years after my meeting with Goldberg, Belgian tax authorities raided Omega's Antwerp office and seized €100 million worth of diamonds. This would evolve into charges related to a multibillion-dollar tax evasion scheme with Omega allegedly at the center of a ring of companies moving diamonds from Angola and Congo through Dubai, Tel Aviv, and Geneva. In 2013, Omega reached a settlement with the Belgian government, paying €150 million in fines and backtaxes.

The UN panel that I served on met people like Goldberg, gun-runners, diamond smugglers and "reformed" killers.

We obtained photographs and first-hand testimony from people directly involved with an aircraft that carried Eastern European arms from Burkina Faso to Liberia, and we listened to radio intercepts of messages between Monrovia, the capital of Liberia, and RUF fighters in Sierra Leone. We met Charles Taylor, by then the President of Liberia, who flatly denied any role in diamond smuggling or anything to do with arming, training or protecting Sierra Leone's RUF rebels. But the evidence against him was incontrovertible, and the connection between diamonds and the on-going war was as clear and as bright as a D-Flawless 10-carater. We asked the Security Council to impose an embargo on Liberian diamonds, and, for good measure, we asked it to place an embargo on all international travel by Taylor, his cabinet, his family, and some of his more dangerous henchmen. It did.

It did because the evidence against Taylor was good, and because the wars in Sierra Leone, Angola, and Congo remained front and center in Security Council consciousness. The wars continued to rage, spilling their banks into Guinea, Côte d'Ivoire, and some of the states bordering the Democratic Republic of Congo. And now the Security Council had its own dog in the fight. In 1999 it converted a small observer team in Sierra Leone into a full-fledged peacekeeping mission, one that was on the way to becoming 6 times the size of the Angolan effort in a country 17 times smaller. This time they aimed to get it right.

The Security Council was not alone in the growing clamor for a diamond control mechanism. The July 2000 G8 Meeting in Okinawa called for "practical approaches to breaking the link between the illicit trade in diamonds and armed conflict." A UN General Assembly Resolution at the end of that year called upon governments and the Kimberley Process to "give urgent and careful consideration to devising effective and pragmatic measures to address the problems of conflict diamonds." The Kimberley Process was under increasing pressure to deliver.

Who writes, wins

The single most important element in reaching an agreement was almost accidental, but it provides a basic lesson for those interested in reaching a solution under difficult negotiating conditions. The chief negotiator for the European Commission, Ton de Vries, had a vague mandate after the fifth meeting to produce a draft proposal dealing with a small part of what was being debated. Instead, he went home and wrote up a complete certification system, codifying what had already been agreed and adding whatever he thought was missing. The outcome, at the sixth meeting in July 2001, was almost universal outrage. How dare he write everything down in such detail, especially things that had not been agreed – things that had not even been discussed?

What happened next, however, is the teachable lesson: who writes, wins. The delegates set to work poring over his draft, focussing at last on a concrete document rather than the unstructured and circular debates of the past. Some of the draft was thrown out, much was amended, and more was added, but after five more tortuous meetings – in London, Luanda, Gaborone, Ottawa, and Pretoria – the delegates gathered for a final negotiating session at Interlaken in the Bernese Oberland of the Swiss Alps in November 2002.

There was a fresh dusting of snow high on the Jungfrau when the delegates took their seats on the last day at the grand old Victoria Hotel. In the town, the NGOs had trudged through the slush from their cheaper hotel down the road, for they, like everyone else at the meeting, had to pay their own way to each of the gatherings. The Taiwanese delegation sat gloomily in the lobby, part of a deal with China that would allow them to participate, but only as part of the World Diamond Council private-sector delegation.

The text on the table, now known as the Kimberley Process Certification Scheme (KPCS), was lengthy. It included elements that some delegates did not like, and it omitted elements that others thought essential. There was something in it, or not in it, to displease almost everyone. But everyone knew that we had gone as far as we could, and, imperfect as it was, it wasn't likely to improve with more debate. It was a question now of take it or wreck it. If the NGOs had walked away, the scheme would likely have died then and there. If they had demanded more, the Russians or the Chinese might have walked out and that too would have killed it. It had to be inclusive; all major diamond-producing, trading, and manufacturing countries had to be part of it, or it couldn't work. Everyone had to agree. Consensus – in this case meaning 100 percent agreement – was required.

The South African Chair asked if anyone in the room disagreed with the proposition on the table, or with a proposal that it should take effect eight weeks hence, on January 1, 2003. For the first time in more than a dozen meetings, involving by now some 45 governments and a wide cross-section of industry and civil society, no hands went up. The KPCS was ready to roll.

CHAPTER FIVE

Regulation

Despite its imperfections, which we will come to in chapter 6, the Kimberley Process Certification Scheme is a remarkable piece of work. It is voluntary in the sense that a country can join or not join, but if it joins, it must abide by certain compulsory undertakings. No member of the KPCS may allow diamonds to be shipped to a non-member; likewise, no diamonds may be imported from a non-member. If a country has a diamond industry, therefore, it must join the KPCS or buy and sell its diamonds outside the system, with other non-members. This was an unviable proposition as some countries soon discovered. Mexico, for example, has no diamond industry per se, but it needs industrial diamonds for its tooling and machining industries. For a couple of years after the KPCS began, Mexican buyers found sellers who were willing to smuggle suitcases of cheap industrial diamonds across borders, but it took only a couple of seizures for them to realize that it would be a lot cheaper if Mexico simply joined the KPCS. It did. So did Indonesia, Cameroon, Bangladesh, Panama, and other countries not present at the outset. All of the big players were in; if these smaller actors wanted to buy or sell diamonds legally, they had to join as well.

During the final negotiations, some countries worried that the KPCS might be seen as a restriction on trade, contravening the spirit and the rules of the World Trade Organization (WTO). After the Interlaken meeting of November 2002, 11 KP member states joined in requesting a general WTO waiver from GATT rules in order to give certainty to the domestic regulations that were being created. The decision, which came in February 2003, recognized "the extraordinary humanitarian nature of this issue and the devastating impact of conflicts fueled by trade in conflict diamonds on the peace, safety and security of people in affected countries, and the systematic and gross human rights violations that have been perpetrated in such conflicts."[1] The three-year waiver was renewed in 2006 for six years and again at the end of 2012 for another six years.

The elegant simplicity of the KPCS lies in the fact that there is no treaty, no agreement to sign, and no UN involvement apart from general encouragement. At Interlaken, the Chair asked for a show of hands from countries willing and able to become members. Those representatives who raised their hands (everyone present) understood that they were committing their government to the creation, passage, and promulgation of legislation and whatever concomitant regulations might be required in order to comply with KPCS standards. It was a rush, but it happened. Russia, the European Union, the United States, China, South Africa, Japan – almost every country present at Interlaken – had within six months passed new legislation and created the supporting regulations required to meet KPCS standards.

The KPCS, therefore, has the force of law in each of its member states. It is not a vague international undertaking or treaty; it is not a list of guidelines or "best practices"; it is the law of the land in more than 50 countries plus the 27 member states of the European Union.

Definitions are important in any agreement, and no less here. The KPCS defines "conflict diamonds" as "rough diamonds used by rebel movements or their allies to finance conflict aimed at undermining legitimate governments, as described in relevant United Nations Security Council

(UNSC) resolutions..."[2] It was also important to define words such as *diamond, export, import, parcel, transit,* and others.

KPCS standards were called "minimum standards" in the sense that they represented a floor beneath which member states agreed not to fall. *Minimum,* however, did not mean *minimal,* and the basic principles are anything but. Each country is required to issue its own Kimberley "certificate" for every parcel of diamonds being exported. The certificate is a government guarantee that the diamonds in the parcel are conflict-free and that the government has an auditable system of internal controls enabling it to track diamonds back to the place where they were mined or, in the case of re-exports, to the point of import. The certificates, mostly produced by security printers and containing a variety of banknote security features, list details of what is in each parcel. Each parcel must be sealed in a tamper-proof manner, and when it arrives at its destination, the government of the recipient country acknowledges receipt by returning a tear-off portion to the sender.

In 2007, police arrested two men attempting to sell diamonds at a Tucson gem show. They could not produce papers showing how the stones had been imported. Later the same year, $10,000 worth of rough diamonds shipped from Spain to a company in Indiana were seized by US Customs agents because they were not accompanied by a Kimberley Certificate. In 2008, a Ugandan was arrested entering Canada with diamonds for which he had no Kimberley documentation. He had purchased the diamonds legally in Uganda and had valid invoices and Ugandan government export documentation. But Uganda is not a member of the KPCS, and the diamonds therefore contravened both KPCS standards and Canada's new KP law. The amount in question was small, but the man was fined nevertheless, and he lost his diamonds.

There is an old mafia saying: "Punish one, teach many." This is a principle of the criminal justice system as well.

These cases and others were widely publicized in diamond circles and helped stakeholders understand what was at risk in attempting to evade the law. A $10,000 mistake is one thing, but most parcels of rough diamonds are worth many times that amount, and losing the goods, along with a fine or prison time, is no laughing matter.

The KPCS included the creation of a statistical database. A special website for KPCS data was created and the data are available to all members and to industry and civil society participants as well. Each member country must submit quarterly trade statistics, and if it is a diamond-producing country, semi-annual production data. It took a couple of years for the system to become operational, a herculean effort that was managed by the government of Canada. The logic behind the concept was clear. With production data, members can see whether a country with known diamond resources is producing more than makes sense. In the absence of obvious new finds, spikes in production can be queried. When it became obvious in 2004, for example, that Congo-Brazzaville was exporting volumes far in excess of its known production capacity, a review mission was set in motion.

The per-carat average price is also an important statistical indicator of wellbeing. Liberia's historical average ran at under $25 per carat during the second half of the twentieth century. When it spiked to more than $700 a carat in 2010, explanations were called for. Trade data provide an additional check. If Guinea reports that it exported a certain weight and value in industrial diamonds to Lebanon, Lebanon should show roughly the same volume, value, and quality coming in, give or take a few points to account for time lags and profit taking. Differences of more than 15 percent one way or the other, however, are an occasion for queries.

Getting the statistical system up and running was like taking a fractious primary school class to the dentist. In an industry so heavily inured in secrecy, objections to transparency flowed like water. "Commercial confidentiality" was one of the first, although a confidentiality problem can only arise in the rare case where a country deals with only a single exporter or importer. Otherwise company data are buried in aggregate numbers and no company names are used. Commercial interests are further safeguarded by the fact that data are not required until six months after the reference period. It *is* required, however, and extracting it in the early days was sometimes like pulling molars.

In the first few years, two other data hurdles had to be overcome. The first was Russia's claim that diamond statistics were a state secret and could never be revealed. The chairmanship of the Kimberley Process rotates annually. South Africa did the heavy lifting up to the end of 2003. Canada chaired in 2004, and Russia was to take the Chair in 2005. When the KPCS was inaugurated, Russia had been given a pass on the submission of data, but late in 2004, a number of African countries said they would veto Russia's upcoming chairmanship if it was unable to achieve standards that were by now being met by everyone else. One of the reasons the Russian delegation had fought the issue so strenuously, it turned out, was that in Russia, civil servants dealing with state secrets warrant a higher pay grade than others, and if diamonds were taken off the state secret list, bonuses might be in jeopardy. In the end, the embarrassing potential of losing the 2005 chairmanship proved more persuasive than state secrets, and Russia finally knuckled under.

An additional problem was whether any of the data should be made available to the public. The industry and a handful of governments fought strenuously against this, arguing that data submitted for KPCS purposes might differ from Customs data appearing elsewhere. Time periods might be different; values and volumes would also, therefore, be different, and uninformed journalists, spotting obvious variances in two sets of public data, might cause grief. It was not until 2008 that the debate ended in favor of publication of some of the data. Today anybody can see quarterly production levels in Tanzania or Australia, for example, and how much the diamonds are worth. Not only is the KP database the best there has ever been on diamond production and trade, much of it is now open to the public, making it an excellent tool for spotting anomalies that should not be there.

Unlike later efforts to halt conflict gold, tantalum, and other minerals from the Congo and its neighbors, the Kimberley Process was not about isolating bad diamonds from good diamonds. It aimed to regulate the entire global trade in rough diamonds. This would have two effects. First, if done well, it would halt conflict diamonds. Second – an important by-product that was unacknowledged in KP negotiations – it would impinge severely on other forms of criminality in the diamond trade.

Early in KP discussions there was talk of alternative approaches. Rough diamonds could be coated with different formulae of mine-specific chemicals and identified later through specially developed spectrography. Bar codes could be etched onto higher-quality rough diamonds before export. Mines could be "fingerprinted" using a technology called laser ablation inductively coupled plasma mass spectrometry (LAICPMS), which can identify invisible impurities unique to the place where a diamond is mined.

These technologies, however, would have added huge costs to each of the millions of diamonds that are produced in a year. LAICPMS would require not just the "fingerprinting" of each mine, but fingerprinting of different depths of each mine because of changing characteristics that developed as diamonds made their way to the surface so many eons ago. On top of that, as with technologies based on marking diamonds in some way, every single rough diamond crossing every single border would have to be checked against whatever database had been established. Aside from the enormity of work required, a question arises about diamonds with no identification. Those with no mark or chemical, or diamonds not listed on the LAICPMS database would not necessarily be conflict diamonds, or even stolen diamonds. They could be nothing more than diamonds without an obvious identity, perhaps from a country that could not afford the identification technology. Apart from the fact that anything added to a diamond can be cut off or changed, any of these systems would require that every single rough diamond in the world be individually identified and cataloged, a proposition of impossible proportions.

The industry opposed ideas like these for an additional reason. Diamonds undergo a basic sorting process in many producing countries before export. The real sorting, however, takes place in the major trading centers of Antwerp, Dubai, London, and Ramat Gan in Israel. Buyers import rough diamonds from a dozen places and sort them according to the requirements of specific customers. Buyer A in New York wants a certain range of size, color and quality, and may in due course receive diamonds that have originated in Sierra Leone, Angola, and Tanzania. Buyer B in Sri Lanka will have other requirements, and may receive stones that originated in Canada, Russia, and Guinea.

Because of the way in which diamonds are mixed, the industry preferred a regulatory system that is essentially based on a wholesaling approach, certifying *parcels* of diamonds at the key choke points of export and import. This means, however, that *all* parcels of rough diamonds – *everywhere* – have to be certified, and, in the process, all smuggling has to be interdicted, not just in places prone to conflict. The KPCS was created to stop conflict diamonds, but its design required it to halt *all* diamond smuggling.

A key to the KPCS was the simplicity of its organization. There would be a Chair, and the Chair would rotate annually. South Africa was first in 2003, Canada came next, then Russia, followed in successive years by Botswana, the European Union, India, Namibia, and others. The Chair's major role is to convene and host two meetings of members each year, one known as the "Plenary" and another to deal with technical issues called the "Intersessional" meeting. Governments alone are "members" of the KPCS, but in recognition of the essential role played in the negotiations by industry and civil society, these were accorded "observer" status. The term is misleading in the sense that industry and civil society organizations continued to play a clear and sometimes powerful role as the KPCS evolved. They participate actively in all Plenary and Intersessional meetings, all working groups and all reviews. When consensus is required, their vote is counted and their "veto," while unlikely to stop a freight train, does have meaning. The "tripartite" nature of the Kimberley Process and the KPCS is one of its unique features, and it is undoubtedly a contributor to the success – and the critique – of the system.

It was agreed at the outset that the KPCS would have no formal secretariat and no budget. Whatever convening and coordination functions it required would be handled by the Chair (at the Chair's own expense), and by a number of working groups (at their members' own expense). While some argued strenuously that the KPCS could not work effectively without – at the very least – a small working secretariat, others disagreed. Russia was particularly adamant that there be no permanent secretariat, arguing that such bodies have a tendency over time to develop policy-making functions and unwarranted authority, with tails too soon wagging dogs.

So a system of working groups was established, including a Statistics Working Group to ensure that data were gathered and analyzed in a timely way. A credentials committee, known as the Participation Committee, was to ensure that member states, or potential members, have the required regulations and standards in place. There is a Working Group on Monitoring – perhaps the most important of all, because it deals with issues of compliance. There is another on rules and procedures. And there is a Technical Working Group to deal with issues such as the harmonization of customs codes that distinguish gem from industrial diamonds, unsorted from sorted, untreated from partially treated, and so on. There is a "Selection" Working Group that acts as a nominating committee for the appointment of the KP Vice-chair, important because this year's Vice-chair will become next year's Chair.

The main advantage of the working group approach, over the creation of a secretariat, is ownership.[3] Any government that wants to be in a working group is welcome, as are industry and civil society participants. This means that issues are tackled by those on the front lines of the KPCS, and if the Monitoring Working Group, say, reaches a conclusion on a difficult subject, it has the automatic backing of its members, obviating the need to second-guess the work of arms'-length bureaucrats. The working groups "meet" by conference call as often as may be needed, and in person during the two annual face-to-face gatherings.

One thing that proved impossible to negotiate prior to January 2003 was a monitoring mechanism. The preamble to the KPCS document "welcomed" self-regulation initiatives and spoke in the vaguest of terms about monitoring. It stated that a monitoring mission might be undertaken in cases of "credible indications of significant non-compliance," but it failed to define *mission, credible, significant,* and *non-compliance.* It would take a year after start-up before a peer review mechanism could be established, one which would see review teams made up, typically, of the representatives of three governments, industry, and civil society doing on-the-ground systems checks of other member countries and determining how well KPCS minimum standards are being applied in action. Although voluntary, it soon became clear that to deliberately avoid a review visit was to invite disapprobation, and the system became part of regular Kimberley Process procedures within a couple of years.

Mark Twain once wrote that the difference between the right word and the almost-right word is the difference between lightning and lightning bug. And so it turned out with the words chosen for the undertakings of the Kimberley Process Certification Scheme. Everyone involved knew that it was imperfect at the beginning, but most believed the imperfections could be ironed out in due course. Some were. But what few could foresee was a KPCS that would ignore member states failing to manage their internal controls convincingly. Nobody imagined that the KP would ignore statistical anomalies as unbelievable as those so dutifully inscribed by Belgian importers during the 1990s. In 2003, it is unlikely that many favored a decision-making process that would lead to deadlock on almost every issue of importance over the next ten years, or a Kimberley Process that would actually condone wholesale diamond smuggling from a member state. They would have been surprised to hear that a former KP Chairman would squeal on an NGO whistle-blower, leading to his imprisonment in one of Africa's most dangerous countries, and that this would be allowed to pass, almost unnoticed. Given how difficult, and yet how civil, the 2000-2 negotiations had been, members probably could not have foreseen the insults and invective that would be hurled about during future Plenary sessions. And surely none could have foreseen corrupt member states committing the most egregious human rights abuse and extrajudicial murder in their diamond industries – and getting off with the very slightest of admonishments.

Despite the apparent checks and balances, the careful planning, the detailed structures, the participatory nature, and its tripartite foundation, the KPCS soon showed itself to be dozy, bureaucratic, and largely apathetic to most of the issues it had been designed to confront. Damning investigative reports by journalists and NGOs would expose KPCS incompetence time and again. Efforts to change and reform the system by well-meaning member states would be met with increasingly strident invective from others, and ever-higher brick walls would be placed in the way of effective action. The power and politics of other agendas trumped the essential purpose of the Kimberley Process, transforming it from a high-minded conflict prevention mechanism into a talk shop full of regulators without the will or the courage to regulate.

Each year after 2003, the outgoing KP Chair would submit a glowing report to the UN General Assembly of all that had been achieved under his or her mandate, ignoring the failure and acrimony. Each year, the Chair would draft a resolution for consideration by the UN General Assembly, and each year the General Assembly would reaffirm its "strong and continuing support for the Kimberley Process and its related certification scheme."[4]

CHAPTER SIX

Power and Politics

Theoretically, power in the Kimberley Process is shared equally among all member states, industry, and civil society. But, as in so many seemingly egalitarian arrangements, some actors are more equal than others. There are two parts to the issue. The first is that member states define their own role within KP working groups and decision-making bodies. Some attend all meetings and participate in all committees. Russia, the United States, the European Union, China, and Canada fall more or less into this category. Others attend all Plenary Meetings and participate in a few committees and review teams. Among these are Angola, Australia, Botswana, Brazil, the DRC, Ghana, India, Israel, South Africa, Switzerland, and the United Arab Emirates. Others show up only for the annual Plenary Meeting, while about 20 countries – representing about a third of the membership – participate in no committees and rarely attend a meeting. Among these are Armenia, Belarus, Japan, Laos, South Korea, Thailand, and Ukraine.

The diamond industry ranges from mining firms at one end of the spectrum to retailers at the other, with very powerful and different interests among the trading and cutting industries in the middle. The NGO coalition as well brings together varied Western and African interests.

The burden for managing the KPCS, therefore, has fallen to those governments, NGOs, and industry players that take an interest, and this makes sense up to a point. The problem is that meetings, committees, and review teams cost money, and each participant must pay its own way. Diamonds are much more important to the economies of Sierra Leone and Guinea than they are to the Russian or American economies, and, by rights, Sierra Leone and Guinea should be at least as active as Russia and the US. But poor countries don't have the personnel or the budgets, and so their participation is determined not so much by interest as by what they can afford. Those without budgets and personnel take a back seat. The same is true for NGOs, and, to some extent, industry.

The second and more debilitating power dynamic is the formal decision-making procedure in the Kimberley Process. The Interlaken document states that "Participants are to reach decisions by consensus." Unlike the more common adversarial approach to decision-making, where disputes are put to a vote and the majority "wins," consensus decision-making seeks general agreement. If it is to work, it requires widespread consultation and negotiation. As a result, consensus decisions may have much more support than those made by majority voting. What happens, however, when "consensus" is taken to mean "unanimity" and there are die-hard holdouts? The Interlaken document said this: "In the event that consensus proves to be impossible, the Chair is to conduct consultations."

That sentence is possibly code for a process understood by those in the know, or it may be diplo-babble meaning nothing at all. As the Kimberley Process began to evolve, however, it came to mean that in the absence of unanimity, there would be no forward movement. In other words consensus became not so much a tool for bringing about generalized agreement as a license to block. Instead of "one man, one vote" as conceived in *Robert's Rules of Order* (and most democracies), the Kimberley Process soon became gummed up in a "one man, one veto" arrangement that would stymie forward movement on almost everything that mattered.

It would have made infinitely more sense to adopt a graduated decision-making system like that of the African Union (AU). When consensus – the AU's default mechanism – fails, the AU turns to a two-thirds majority. Procedural matters, including the question of whether a matter is one of procedure or not, are decided by a simple majority.

What follows are vignettes describing the major challenges faced by the Kimberley Process in its first decade. They are gathered under the four broad organizing principles of the KPCS. The first was a requirement that each country bring its diamond industry under a commonly agreed set of internal controls. Without clear knowledge of where a diamond was mined or how it entered the country, a government would be unable to give it any kind of reliable certification. Internal controls, therefore, are the foundation of Kimberley effectiveness and credibility.

The second principle had to do with smuggling. As noted before, in order to create a simple and relatively inexpensive system for the interdiction of *some* diamonds – notably those fueling conflict – it was agreed that *all* diamonds would be encompassed under the KPCS, and that all diamond smuggling would be halted. Many countries had no proximity to the diamond wars, but diamonds are as portable as they are valuable, and it was deemed easier and cheaper to take a global approach to the certification of national systems and parcels of diamonds – a wholesale methodology – than to attempt the certification of each and every diamond from troubled areas. This was an immutable organizing principle. It did not apply to some diamonds and some countries. For the system to work, it had to apply to all.

The third organizational principle had to do with statistics. A reliable and up-to-date database on diamond production and trade was essential to the triangulation and verification of diamond shipments. Today the KP's formidable database is the strongest tool it has for rooting out smuggling, money laundering, and some of the other tricks of a trade that had allowed and even encouraged conflict diamonds. But detection tools are only as good as a system's willingness to use them, and in this the Kimberley Process has stumbled repeatedly.

The fourth organizing principle had to do with the definition of a conflict diamond – one used "by rebel movements or their allies to finance conflict aimed at undermining legitimate governments." This seemed adequate in 2003 when few envisaged a "legitimate government" using untoward force, human rights abuse, and extra-judicial killing as a means of enforcing Kimberley Process internal controls. When the worst happened, however, the Kimberley Process proved more adept at tearing itself apart than at dealing with the issue.

Compliance: internal controls

Republic of Congo

It was not always thus. One of the first tests of the Kimberley Process was posed by the Republic of Congo (ROC), which had long been a transit hub for diamonds smuggled across the River Congo from the Democratic Republic of Congo (DRC). When the 2003 statistics for ROC were examined, they showed large export numbers against almost no production, and nothing at all in the import column. The Canadian Chair of the Kimberley Process, Tim Martin, visited Brazzaville and persuaded the government to entertain a KP review mission. A KP team arrived in May 2004 and was told that the diamonds had certainly been mined in ROC. The team was welcome to inspect the mining areas, but unfortunately the roads were terrible, and anyway, there was rebel activity in the area and the team's safety could not be assured. Undeterred, Canada chartered a Fokker F-27, and with ROC government officials aboard, the team flew up and down the countryside searching in vain for the elusive mines.

The team's report was unequivocal. The Republic of Congo had virtually no internal controls and it could not account for its voluminous exports, which far exceeded known production capacity. The Chair issued a report saying: "The Republic of Congo therefore fails to meet the minimum requirements of compliance for the implementation of the KPCS," and, "given the large volumes of diamonds flowing through the Republic of Congo, urgent and decisive action is required to address the situation . . . In light of this situation and the risk it presents to the integrity and credibility of the KPCS, the Chair issues a revised list of Kimberley Process Participants from which the Republic of Congo has been removed."[1]

In other words, the Republic of Congo was expelled from the Kimberley Process. The evidence of "non-compliance" was as incontrovertible as it was damning. Tim Martin had consulted widely within the KP membership but he never put the decision to a vote. He assumed he had consensus and didn't bother waiting for a meeting where vacillating hands might have started going up. He did the right thing and it might well have set an important precedent. But it didn't. Worse displays of non-compliance were to come, but no subsequent Chair had Martin's combination of good sense, purpose and resolve. The Republic of Congo would be the KP's first, and perhaps last, expulsion for bad behavior.

Democratic Republic of Congo

The expulsion of the Republic of Congo helped to stop the hemorrhage of diamonds into that country from the Democratic Republic of Congo, but it did nothing to improve the DRC's own internal controls. The DRC's willingness to abide by Kimberley Process standards may have been strong when the KPCS began, but thanks to generations of government predation and decay dating back to colonial times, its capacity was feeble.

Solid internal controls are, nevertheless, the bedrock of KPCS effectiveness. If a country cannot guarantee the origin of the diamonds it is certifying for export, then the entire purpose of the Kimberley Process is defeated at the very beginning of the diamond pipeline. And if the KPCS cannot be made effective in countries that have suffered at the sharp end of conflict diamonds, there probably isn't much point in trying to make it work elsewhere.

The DRC, therefore, along with the other victims of conflict diamonds – Angola, Sierra Leone, Liberia, and Côte d'Ivoire – was a kind of proving ground for the KPCS. But during the first decade of KPCS implementation, The DRC turned out to be the opposite: a poster child, in a sense, for how internal controls should not work.

The Kimberley Process sent a peer review team to the DRC in October 2004 and its report was sympathetic to the challenges faced by the government:

> The size of the country, the spread of (largely alluvial) diamond production over several provinces and the large number of artisanal diggers (of whom there are estimated to be between 700,000 and one million, many of whom are probably unlicensed) would present any diamond-producing country with an enormous challenge. The challenges of rebuilding the central administration after years of civil war further complicate the situation.[2]

The team heard accolades for the removal of the Republic of Congo from the Kimberley Process, and heard estimates that as much as 40 percent of DRC production had been smuggled out before the expulsion. That aside, officials estimated that 20 percent of the DRC's production was still leaving under the radar. The actual volume of smuggling might, in fact, have been higher or lower, because the team found that "the production of artisanal miners is not at present verified in a systematic manner." Since 100 percent of all diamond mining in the DRC was being produced by artisanal miners, this was a polite way of saying that the government's internal controls – such as existed – were not working. Buried in a series of recommendations about building the capacity of government departments was the crux of the report: "The DRC should report production figures semi-annually as requested by the KPCS."

This was a bit like recommending that daffodils should bloom in the Arctic. But, having made that recommendation and 21 others, the team submitted its report, the government of the DRC acknowledged receipt, and life in the country's diamond sector moved on without further attention from the Kimberley Process. Over the next five years, a series of official United Nations reports, NGO investigations, and media stories indicated that there was little change. Certainly there was no follow-up from the Kimberley Process. And so the problems festered until March 2009, when a second KP review team visited the DRC.

This time the report was more direct and more critical:

> The team noted that there continues to be very little or no oversight of diamonds between the time they are extracted and the time they are purchased by a *comptoir* [buying house]. There seems to have been little, if any, progress in registering artisanal diggers and traders, and there does not seem to be a system in place to collect the reports and trading data that these actors are required by law to submit. It also seems that, while government agents do attempt to collect production statistics, they are not able to collect statistics on *all* production, meaning that the production statistics submitted to the KP are probably not reflective of what is actually produced.[3]

The 2009 report, running to more than 100 pages, was meticulous and comprehensive, and contained dozens of detailed recommendations. Once again, the report was received, noted, and shelved. The Kimberley Process has no mandate to insist that recommendations be implemented, and it has no formal mechanism for follow-up.

Shortly after the release of the second KP report, Partnership Africa Canada produced a study of its own, one in a series examining the diamond industries in the DRC, Angola, and Sierra Leone. PAC was considerably less polite:

> Congolese authorities have never put more than the rudiments of an internal control system in place. There are no systems for gauging diamond production or tracking internal sales. Diamonds are registered only as they enter the large *comptoirs* or buying houses – located mostly in the capital Kinshasa – where no questions are asked and no identification is required. The Kimberley Process bears a large share of responsibility for this situation. Year after year, the KP has been advised of the essential weakness of the Congo's internal controls by Partnership Africa Canada, other NGOs, the United Nations and its own monitors. Year after year, the KP has chosen to do nothing, and to allow the Congo to do nothing. As a result, the DRC's network of *comptoirs* is the world's most effective system for laundering conflict, illicit and clandestine diamonds . . .
>
> The 2004 KP review of the DRC did recommend ways that [the] loopholes could be closed, but nothing was done by the DRC government, and there was no follow-up by the Kimberley Process. The mid-2009 KP review of the DRC goes into this problem in detail and makes similar recommendations. But recommendations + inaction = zero. The KP, the DRC government and the diamond industry at large need to take a much more proactive approach to this very large problem.[4]

Mostly, they did not.

Brazil

Partnership Africa Canada demonstrated in another report that problems with internal controls were not unique to the DRC or to Africa. In response to pervasive industry rumors of rampant fraud in the Brazilian diamond industry, PAC produced an investigative report that enraged Brazilian authorities. And no wonder, given its provocative title: *The Failure of Good Intentions: Fraud, Theft and Murder in the Brazilian Diamond Industry*. Written in 2005 by Shawn Blore, an intrepid Canadian journalist living in São Paulo, the report told a story that was quite different from that of the Congo. In Brazil, the paperwork was good. Each Brazilian Kimberley Certificate issued by the government was supported by a paper trail leading back to legal mining claims. This in itself was actually somewhat doubtful, however, not least because an estimated 80-90 percent of Brazil's diamond production was in the hands of unlicensed *garimpeiros*. There was the additional problem of illegally mined diamonds emanating from the Cinta Larga Indigenous Area in the remote northwest of the country. In 2004, 29 non-Indian diamond diggers had been murdered on the reserve, which was now under an ineffective ban on all outsiders and all diamond mining.

Blore decided to put the Brazilian system of internal controls to a simple test. He would visit the place where 6,876 carats, according to Kimberley Certificate No. 64, were mined. He was more than a little interested, firstly because the miner in question had apparently sold the entire 6,876-carat lot only days after his mining claims had been issued. The miner sold the diamonds to a company in Belo Horizonte for $261,000 and after three weeks they were flipped to another company for $982,000. That seemed like an unusually large mark-up, and Blore's suspicions were further piqued by the fact that the new buyer, Hassan Ahmad, was a Sierra Leonean citizen of Lebanese descent, resident in Brazil for only five years. In fact, he had close family ties to a Lebanese diamond company in the DRC that was under investigation by Belgian authorities.

Then something even more remarkable happened. Ahmad sold the diamonds to a company in Dubai for $2.67 million, even though the stated value on KP Certificate 64 was $983,000 only a fraction higher than Ahmad's own purchase price. The "anomalies" – as such things are often called in Kimberley Process parlance – tripped over one another, and there was more. When he got to the mine site – dutifully detailed in the application for export – Blore found that nothing had disturbed the ground there for years. And the miner who had supposedly produced and sold the diamonds, one Fabio Tadeu Dias de Oliveira, a small-time crook, had been dead for three years when the first sale supposedly took place.[5]

The PAC report was a devastating critique of Brazilian controls, not least because 34 out of the 61 KP certificates issued by Brazilian authorities between 2003 and 2005 represented diamonds exported by Hassan Ahmad, a man with direct ties to countries wracked by conflict diamonds. The Kimberley Process could have done a number of things when presented with the PAC report. The least would have been to mount an immediate investigative review of its own. Instead, it did nothing.

The Brazilian Ambassador to Britain wrote to the *Financial Times*, which had carried the story, denouncing PAC. The Secretary of Brazil's Ministry of Mines and Energy wrote to the Chair of the Kimberley Process saying that that PAC had "distorted information," had written in "ironic and sneer language," had ignored the Brazilian government's tremendous efforts in the informal mineral sector, and had intruded into Brazil's constitutional process. Further, it had questioned the "objectives and honesty" of several named companies in a report that was "very harmful to the KPCS." Finally, PAC's behavior represented "external interference in Brazilian mineral policies."[6]

Officials of the National Department of Mineral Production (DNPM), which is responsible for Kimberley Process oversight in Brazil, made a presentation at the KP's June 2005 Intersessional Meeting, again attacking PAC's veracity. And five months later, at the November 2005 KP Plenary Meeting in Moscow, the DNPM presented a report entitled "Site Inspection Report; Kimberley Certificate No. 64." This report, written by DNPM Engineer Geologist Luiz Eduardo Machado de Castro and presented to the entire KP Plenary Meeting in detailed PowerPoint, attacked PAC again, this time dealing with and denying the specific allegations of fraud in connection with KP Certificate 64.

Throughout, the Kimberley Process behaved like a small nocturnal animal caught in the headlights of an oncoming vehicle. The Brazilian Federal Police, however, knew what to do about headlights. They had read the PAC report and, while the escalating DNPM protests served to paralyze the Kimberley Process, the police were quietly running their own investigation. In February 2006, they moved. DNPM's own Machado de Castro was arrested in connection with the issuance of false KP certificates, and arrest warrants were issued for nine others, including Hassan Ahmad.

This story has a positive ending, no thanks to the Kimberley Process. The DNPM shut down all Brazilian diamond exports for a year while it reviewed and revised its systems and regulations. When Brazil returned to active participation in the Kimberley Process, its controls were tougher and more effective, and in open session at the KP Plenary Meeting of 2006, Brazil formally apologized to Partnership Africa Canada for what had happened.

The diamonds that had been exported under false certificates were clearly not what they purported to be, but what exactly were they? They might well have been from the Cinta Larga reserve. They might have been produced by illegal Brazilian diggers. But, given Hassan Ahmad's connections with Africa, there is every possibility that he was laundering or planned to launder Sierra Leonean or Congolese diamonds. Not only that: the huge variations in invoice prices suggest that industrial-strength money laundering was taking place as well.

The overarching lesson is that, because of their great value, illicit diamonds are likely to seek the path of least resistance, and the Brazilian episode is a demonstration of the logic behind the inclusion of *all* diamond-producing countries in the Kimberley Process. It demonstrates the need for three additional things that the KP did not have in 2005 and does not have to this day: its own research capacity, rigorous independent monitoring, and a willingness to act swiftly when problems are identified.

Compliance: smuggling

Venezuela

The Kimberley Process became paralyzed over Brazil, in part because politics were allowed to trump common sense. Russia, the Chair of the Kimberley Process during 2005, was unwilling to embarrass a member state on the basis of research conducted by an NGO. And confronted by a withering offensive from the Brazilian government, most KP members decided individually – and therefore collectively in a consensus-based mechanism – that inaction should be the default position. Despite the seminal lessons of the Brazilian case, the same deference to bluster, the same disregard for facts, and the same inaction would plague the Kimberley Process as it grappled with subsequent problems in Venezuela.

By world standards, Venezuela's diamond output is small

– about 150,000 carats a year, worth perhaps $15 million, or perhaps double that amount. There are no large companies involved and all of the mining, which takes place mainly south of the Orinoco River in lands bordering Guyana and Brazil, is done by artisanal miners. Venezuela joined the Kimberley Process in 2003, passing the requisite legislation in May of that year. Soon thereafter, however, Venezuela dropped off the Kimberley radar. A tiny handful of KP certificates covering 33,000 carats was issued in 2003 and 2004, but after that, nothing was heard from Venezuela – no certificates, no exports, none of the required production and trade data: nothing. During 2005 and 2006, the KP made various efforts to regain contact. Queries were sent from different working groups and from the Chair. The embassies of KP member states made representations to the Venezuelan government in Caracas, asking for an explanation. Still nothing.

Clearly, Venezuela was in a state of serious non-compliance, but the Kimberley Process was stymied. According to the rules, it should have sent a review team to find out what was happening, but this could only be done with permission of the host government, and the KP couldn't get a response from Venezuela to anything. Even diplomatic *démarches* resulted in silence. The alternative, to "drop Venezuela from the list," was discussed, but several KP member governments argued that expulsion – especially in the face of Venezuelan silence – was too drastic. Once again, Partnership Africa Canada stepped into the breach, sending Shawn Blore to find out what was going on.

Blore traveled extensively in Venezuela's diamond mining areas. He spoke with government officials and he spent time in Santa Elena de Uairén, a diamond-trading town on the Brazilian border, two hours by road from Boa Vista, capital of Brazil's Roraima state. He went to Boa Vista as well and met diamond buyers there. In November 2006, PAC published

Blore's report, *The Lost World: Diamond Mining and Smuggling in Venezuela*. The title was an allusion to Arthur Conan Doyle's *The Lost World*, a 1912 novel set in Venezuela's remote diamond lands, and it was a metaphor for the Kimberley Process as well.

Two things were going on in Venezuela. The first was that, in 2005, the mining portfolio of the Ministry of Mines and Energy had been spun off into the Ministry of Basic Industry and Mining. In the reorganization, Kimberley Process responsibilities simply went missing. Nobody was given authority to issue KP certificates and nobody had a mandate to respond to KP queries. Compared with other priorities during the chaotic political and economic changes taking place in Venezuela at the time, diamonds were insignificant. And in the diamond mining areas, a grand total of three individuals equipped with a single vehicle were expected to supervise vast numbers of miners across huge swathes of territory. The upshot was that nobody in the Venezuelan government knew what was going on, and nobody was authorized to do anything.

The bureaucratic confusion suited those in Venezuela's diamond industry, where the second issue had to do with the cost of doing business. Currency exchange controls introduced in 2003 had created a 25 percent spread between the official and black-market rates for the Venezuelan bolívar, and a requirement that legally exported diamonds be exchanged for US dollars helped to push the trade underground. In addition, Venezuela had introduced a 14 percent VAT in 1999, which meant that diamond traders operating legally had an additional burden. This was made worse in 2003 by the introduction of a "Zero Evasion Plan" which contributed to the situation that Blore found in 2006: a reality characterized by 100 percent evasion.

That reality was available for anyone to see in Santa Elena and in Ciudad Bolívar, capital of Bolívar State. Brazilian and

Guyanese diamond buyers were operating in plain sight, easily found just around the corner or in unmarked upstairs offices. Thinking their visitor was a potential client, most spoke openly to Blore about how diamonds were moved across borders, readily transformed into Brazilian and Guyanese diamonds and certified under KP regulations in those countries. To help prove the point, the decline in Venezuelan diamond exports during the early 2000s correlated nicely with a spike in Guyanese production.

The issue could have been resolved in one of two ways. Venezuela could have fixed the problem, or the KP could have expelled Venezuela. Neither happened. In 2007, Venezuelan government delegates at last showed up at a Kimberley meeting. They denounced Partnership Africa Canada but they did acknowledge problems. They promised to fix them, and under duress agreed that the Kimberley Process could send a review team to ascertain the truth – against what they claimed were PAC's untruths. The KP Chair that year was the European Union, and the Chairman, Karel Kovanda, expressed his "delight" at Venezuela's return.[7] He was delighted, no doubt, because the Venezuelan problem contradicted the myth of an unblemished KPCS which each Chair liked to present at the UN General Assembly at the end of his or her tenure, and because clutching at straws was more agreeable than grasping the nettle. Like others to come, the EU Chair aimed to get past the end of his year without a diplomatic brawl, and to pass any incipient crisis on to his successor.

The reason for inaction and deadlock over Venezuela had nothing to do with KP rules, conflict prevention, doubt about the facts or anything to do with diamonds. The problem was one of geopolitics. Venezuelan President Hugo Chavez was a hero for some in the developing world, and while a few members of the KP were prepared to get tough on the diamond front, others were not. In addition, Venezuela was spending billions of dollars on Russian military equipment and had partnered with Russian and Chinese firms on oil, gas, and other mining operations. Plus, Venezuela has more known oil reserves than any other country on earth, which is a big help in winning friends and influencing people.

So in the absence of consensus, the Kimberley Process badly needed a face-saving solution. Instead, what it got was face-slapping. Venezuela stalled on allowing a KP review mission until October 2008, two full years after the contretemps began and four years after the country had stopped issuing KP certificates. Venezuela refused to allow a civil society representative on the team and denied the team access to any of the diamond areas. The team put the bravest face it could on the matter, working out a consensus report that essentially accepted Venezuelan promises to do better soon. A polite fiction was "welcomed" by the Kimberley Process when Venezuela offered to "self-suspend" participation in the Kimberley Process for two years while it undertook to reform its systems. As part of the deal, Venezuela promised that any diamonds mined during this period would be "stockpiled" (as they claimed had been the case over the previous four years) and that there would be no exports.

Nothing, in fact, changed. Shawn Blore visited Venezuela again in 2009 to find that mining carried on as before and the busy trade in diamonds on the Brazilian border continued. The new PAC report achieved little, however. In a November 2009 letter to that year's Chair of the Kimberley Process, Venezuela's "Minister of the Popular Power for Mining and Basic Industries" said "We ratify our willingness to comply with the KPCS once our voluntary suspension period has expired." He took the opportunity, however, to warn against the "unacceptable" and "flagrant violations of our sovereignty" by missions carried out by "some organisations" in 2006 and 2009. He "warned" that rulings against this sort of activity would be "strictly enforced."[8] Obviously Venezuela did not appreciate visitors showing up unannounced in the diamond mining areas.

Over the next three years, civil society participants in the Kimberley Process and a handful of governments would continue to press for action. Venezuelan envoys would appear occasionally at KP meetings to denounce their critics, to ask for more time, and to make more promises about future compliance. They were always given another chance, but, as before, reports failed to arrive, diamonds continued to flow, and while deadlines were sometimes set for a final showdown, it never happened. There was no consensus.

In August 2012, a *Time Magazine* reporter followed the same trail that Shawn Blore had blazed in 2006 and 2009:

> In an office in Santa Elena, deep in the Venezuelan jungle bordering Brazil and Guyana, a diamond trader inspects a rough gem under his magnifying glass. Surrounded by precious minerals, stuffed tarantulas and a sprawling anaconda skin pinned to the wall, he takes calls from men who work in the rowdy clandestine mines nearby and bring him the precious stones. From there, a broker will traffic the diamonds into Guyana, where they'll receive falsified certificates that they were legally mined and marketed. Many will end up in commercial hubs like New York, Tel Aviv and Antwerp.
>
> And the entire journey will flout the Kimberley Process, a decade-old, U.N.-mandated international agreement to curtail rampant global diamond smuggling. Venezuela, a major diamond producer, is a KP member but voluntarily removed itself as an active participant in 2008 after being widely accused of ignoring the pact's mission to regulate diamond production and commercialization. "There is no control at all," says the Santa Elena trader, who asked not to be identified. The KP, as a result, is considering expelling Venezuela: the U.S., which chairs the KP for 2012, this summer delivered an ultimatum to Venezuelan authorities to demonstrate compliance or lose membership altogether.[9]

It sounded good, but the best that could be achieved at the Plenary Meeting in November 2012 was another promise from Venezuela and a vague KP statement that "If these steps are not taken by April 1, 2013, the Plenary regret[s] that appropriate actions will be taken, which may ultimately lead to Venezuela being removed from the KP."[10] Unsurprisingly, April Fool's Day 2013 came and went without "appropriate action" – or action of any other kind.

There are obviously good reasons for caution in the expulsion of a diamond-producing country from the Kimberley Process. Unregulated diamonds from an outlier can only encourage and spread illegality further into the system. Every diamond being smuggled out of a country with poor controls represents a diamond being smuggled into another country where there is obviously a concomitant regulatory problem. In other words, it takes two to tango. The preferred solution in this case was for the government of Venezuela to enforce its laws and keep its promises.

In the absence of that, however, the diamonds were getting out into the world anyway, and the Kimberley Process had, and still has, an obligation to get tough if it wants to preserve the integrity of the system and of diamonds. When it fails to do so, as in this case, it opens the door to similar behavior elsewhere. Why should other countries enforce costly rules and regulations, submit detailed reports, and attend unproductive meetings in faraway places if these have no meaning, and if there are no consequences to ignoring the entire system?

This was made clear to Blore by Guyanese officials in Georgetown. When he first visited, they were seriously concerned about losing their KP membership. "Diamonds are a serious business in Guyana," Blore writes:

> and a ban would have been disastrous. So they were keen to fix any problems. Fast-forward three years: after watching the KP and Venezuela, Guyana now worries not a whit.

> Which is to say that, though the volume of Venezuelan diamonds was insignificant, the damage to the KP's reputation – within its own membership – was vast and immeasurable. After Venezuela, not even Participant governments took the KP seriously anymore.[11]

The most damning critique of the Kimberley Process in relation to Venezuela is not that it did nothing of any substance for seven years, or that it allowed itself to be led down a winding garden path by a feckless and apparently incompetent government. It wasn't even that it could not achieve consensus in a debate polarized by larger issues, although it did demonstrate – just as in any policing arrangement – that consensus decision making and law enforcement make poor bedfellows. The most damning critique is that a regulatory body – established to prevent conflict diamonds through a proscription on smuggling – actively and knowingly condoned the thing it was established to prevent.

India

If rough diamonds are being smuggled out of a country like Venezuela, they have to find a buyer somewhere. There are two generic possibilities. One is that they are smuggled into a neighboring country, like Guyana or Brazil, where they are passed off as locally produced goods. Presented for export, they receive a Guyanese or Brazilian KP certificate and, lightly laundered, are soon on their way. The other possibility is to avoid the Kimberley Process entirely and take the diamonds straight to a cutting and polishing factory somewhere. Cutting and polishing is done in many countries, but an estimated 85 percent of it by volume is done today in India. So if you can get the goods past India's Directorate of Revenue Intelligence (DRI), there are no further hurdles.

That is because the Kimberley Process takes no interest in what enters or leaves a cutting and polishing factory, and this is one of its greatest blind spots. Polishing companies are not obliged to keep KP-related records, and nobody asks them about the origin of their goods. When diamonds are offered at the front door, or at the Bharat Diamond Bourse in Mumbai – the world's largest – or on a backstreet in Surat's Mahidharpura diamond district, they are deemed to be legal because they are presumed to have entered the country with full and proper oversight by the government. But a lot of people enter India every day and fail to declare what's in their luggage. As a result, the DRI has made diamond seizures worth millions of dollars in recent years, *after* they have entered the country. A senior DRI officer told the *Times of India* in August 2011, for example, that the Directorate had picked up 58,500 carats in the previous four months alone – from Congo, Zimbabwe and elsewhere.[12] Some of the diamonds that are seized by the DRI turn out to be legitimate, and once the paper work is more carefully examined, they are released. Others are not. And still others are simply not seen by anyone except the less scrupulous of the 3,000 Indian factories where diamonds are processed.

In February 2013, between $50 million and $350 million in rough and polished diamonds was hijacked by armed robbers at Brussels Airport.[13] If the Kimberley Process had a watching brief on the cutting and polishing sector, thieves would have to take care in laundering such a large quantity of rough. But it doesn't.

It would be simple enough to include cutting and polishing houses in the KPCS. All owners know precisely what enters and leaves their factory. Everything is carefully identified, weighed, and watched. Even the diamond dust that results from polishing is weighed and sold. This is as true in New York as it is in Tel Aviv or Surat. It would not be a major effort to require these companies to produce invoices relating to a KP-authorized import, and this would not be difficult to police on a spot-check basis. Although this has been proposed many times to the Kimberley Process, the idea has never reached first base. This issue is important everywhere diamonds are cut and polished. But, because of the size of the business in India, there it is essential. And absent. As one writer put it, "India is where the Kimberley Process goes to die."[14]

Compliance: statistics

The Kimberley Process statistical database is one of its most powerful tools for monitoring the international movement of rough diamonds. Each member state must submit quarterly trade data and semi-annual production data, along with a breakdown between gem- and industrial-quality diamonds, and details of in- and outbound shipments by country. All of this is captured on a special KP statistics website, some of which is open to public scrutiny.

The site – created and maintained by the government of Canada until 2009, and then by the United States government – requires each participating country to enter its own data. This means that the workload for the KP Statistics Working Group and its Chair is limited primarily to data analysis. The data allow various kinds of study. If Country A reports that it shipped 20,000 carats worth $3 million to Country B, something akin to these numbers should appear in the import statistics for Country B. There will be a few discrepancies if entries are made in different reporting periods, but they will even out over time. If there are major inconsistencies, however, they can be queried and explanations sought.

In addition, production figures should correlate with a country's known production capacity. This is what caused the Republic of Congo such grief in 2004 when it exported far more diamonds than it produced. Because each country enters its own data, incorrect or false figures can, of course, be submitted, but if there is expert analysis, major anomalies can be caught and queried.

In fact, the analysis does not have to be "expert"; it requires basic knowledge of diamond production and trade, a calculator, time and some common sense. The agreed standard for the triggering of a query is a change in data of more than 15 percent from one reporting period to the next. Because the Kimberley Process resisted the creation of a permanent secretariat to handle such matters, data analysis falls to the members of the Statistics Working Group. Like other working groups, this is composed of representatives of a dozen governments, plus one each from industry and civil society. The task is divided up, with each member reviewing four or five sets of data. The results, however, vary in depth and quality, and queries can be ignored, or answered in ways that provide little, if any, meaningful explanation.

Guinea

Guinea provides an example of a country whose statistics, by 2008, made no sense, and where urgent remedial action was required. Among the oddities in the data was a 554 percent increase in diamond production over two years, and a drop in the per-carat average value from $84 to only $17. By weight, 63 percent of Guinea's production was going to Lebanon, most of it listed as industrial diamonds with an average per-carat value of only $1.97. There were two things worth noting here. The first was that $1.97 – low even for industrial diamonds which usually sell for about $7 to $10 per carat – was a tiny per-carat fraction of the value of Guinea's industrial diamond exports to other countries. Why did Lebanon want such poor-quality industrial goods? The second was that Lebanese statistics showed exports of gem-quality rough far in excess of what was being imported. It appeared that large amounts of Guinean diamonds were, for some reason, being sold at bargain basement prices to buyers in Lebanon who knew what Guineans did not – that many of these "industrial" diamonds were actually of gem quality and were enjoying a 500 percent mark-up on re-export.

The Kimberley Process conducted one of its least successful peer reviews that year in Guinea. For some reason, Guinea was a popular destination; in addition to one team member each from industry and civil society, seven government representatives, instead of the usual three, flew to Conakry for the exercise. And except for a flight up-country and a single hour on the ground in the diamond mining area, in Conakry they stayed. It took almost a year for the team to produce a report that ignored charges from Guinean diamond dealers of governmental corruption and mismanagement. And the report ignored the statistical problems as well. It hardly mattered, however, because between the team visit and the issuance of its report, a coup took place and almost nothing the team wrote had any relevance to the new reality in Guinea.

In 2009 the authoritative industry publication *Diamond Intelligence Briefs* weighed in, saying that the KP's slow reaction on Guinean statistics looked like "wilful ignorance" and that the Kimberley Process was ignoring evidence from anti-money laundering agencies demonstrating that there was in the numbers, "prima facie evidence of unadulterated money laundering" and perhaps more.[15] "More" referred to an allegation that Hezbollah operatives in Lebanon were directly involved in fraud, money laundering, and diamond smuggling.

There were other possibilities in the strange Guinean numbers. Between 2003 and 2012, the only conflict diamonds recognized as such by the Kimberley Process were emanating from Côte d'Ivoire. Using satellite photography, the KP kept close tabs on the steady growth in mining behind rebel lines. But it was never able to discover how the diamonds were leaving the country, and it was never able to stop them.

A quick look at a map of West Africa – no satellite required – shows that the rebel-controlled areas in Côte d'Ivoire butted up against the Guinean border. Given the corruption and the lack of internal controls in Guinea, a suspicious mind might have concluded that Guinea provided an obvious and wide-open exit route for Ivorian conflict diamonds.

The Kimberley Process did take note of external reports and at last agreed that Guinean statistics made no sense. It created an action plan that was accepted by Guinea at the 2009 KP Plenary Meeting, requiring much more detail on production, better controls and better internal monitoring. The US Geological Survey assisted over the next three years, and at the 2012 Plenary Meeting the KP pronounced itself satisfied that the problems had been solved. It took almost three years from the 2009 Plenary to get a review team into Lebanon, however, and more than a year more for the team to squeeze out a report concluding that Lebanon was actually meeting KPCS minimum standards. The report's few paragraphs on the mysterious statistics are superficial in content and bewildered in tone. Lebanon is admonished to do better, bringing to mind the old Persian proverb: "The dogs may bark but the caravan moves on."

The Guinea–Lebanon case is especially egregious. Diamond laundering was clearly taking place with government collusion if not involvement. A country with terrorist connections was involved, and the most likely source of at least some of the diamonds was rebel-held mines in an African country at war. This is what the KP was created to address and yet it took years for it to act and, in the end, it produced nothing more than a squib.

Liberia

The war in Sierra Leone ended in 2002, and with it the meddling of Liberian President Charles Taylor. Taylor himself was forced to resign in 2003, going into a long exile in Nigeria before being arrested and placed on trial in a war crimes court in The Hague in 2007. Sierra Leone had taken part in the discussions that led to the creation of the Kimberley Process and was a charter member, but Liberia remained outside until UN sanctions against its diamonds were finally lifted in 2007. The lifting was accompanied by an intensive program of assistance from the Kimberley Process, the United States, and several private-sector companies, aimed at putting in place Liberia's first-ever serious regulatory system for diamonds. It included a network of government inspection stations, a detailed system of internal controls, and a government valuation office in Monrovia to certify the diamonds presented for export.

It was probably one of the better diamond control mechanisms in Africa – on paper. And it worked fairly well for a couple of years. One of the problems Liberia faced, however – apart from a ruined economy and a need to rebuild its government from the ground up – was a very small diamond sector requiring an administrative structure that could barely pay for itself. Liberia's historical diamond-production figures have always been something of a guesstimate because of the smuggling from Sierra Leone that was so rampant in the 1950s and later. But estimates place the volume at something between 25,000 and 75,000 carats per annum of very low-quality diamonds, worth $25-30 a carat.[16] Using the most generous figures, this would yield about $2.25 million per annum. The export tax on this – typically 3 percent – would yield less than $70,000. Even with exploration, mining, and export licenses, the government's gross annual income from diamonds would probably not exceed $250,000. Against this, it was expected to operate a chain of inspection offices, run an export service, print export certificates, send officials to KP meetings, and provide a variety of reports and statistics. This made sense when the world was watching as closely as it was in 2007, and when there was donor assistance, but when that ended, diamonds began slipping down the government's list of priorities.

This, in fact, is a basic problem for the KPCS: its cost to poor countries where diamonds do not bring in much revenue. This is probably why it took Cameroon a full ten years to join and why Gabon is still not a member, preferring to allow its diamonds to make their own quiet way into the (under) world without the cost and bother of a KP membership. And it is undoubtedly a reason for the minimal participation of many other countries. The cost is simply too high, a problem never addressed by the Kimberley Process as a whole. The European Union model could have provided a possible solution. The EU has 27 member states, but only 6 – Belgium, Bulgaria, Romania, the Czech Republic, the UK, and Germany – can import and export rough diamonds. Others must run their imports and exports through one of these countries. This is not an issue of capacity or sovereignty, nor does it imply criticism. If more countries want the privilege, they can have it in return for the systems they will have to develop. The current arrangement is simply a matter of economics and convenience. The same sort of idea might have been developed for smaller producers in Africa, but it was not.

The statistics underlying Liberia's first full year as a KP member should have occasioned some raised eyebrows. Its stated production figures for 2008 showed 47,006 carats mined, and exactly the same number exported. This suggests that the production figures were simply imputed from what was offered for export, and that, despite the installation of government field inspection offices, the authorities had no idea what had actually been mined. Of greater interest was the per-carat average value of the exports. At $210 per carat, Liberia had suddenly leapt into number three position behind Lesotho and Namibia for having the world's most valuable diamonds. In 2009, exports dropped by 40 percent but the per-carat average value increased by 57 percent to $329. And in 2010 the per-carat average was off the charts at $732 – three times the Sierra Leone value, four times higher than Canada and South Africa, and six times more than Botswana.

Only Lesotho was higher, but in Lesotho there is an explanation: the country has long been a source of large, high-quality diamonds. The Letseng Mine is the seventh-largest in the world and it has produced 3 of the world's top 20 rough diamonds. The 603-carat Lesotho Promise was found in 2006, the 493-carat Letseng Legacy in 2007, and the 478-carat Leseli la Letseng in 2008. The Lesotho Promise sold for $12.4 million, representing over $20,000 per carat, a figure that does a lot to raise averages. Although Liberia would claim that some unusually good finds were responsible for its high averages, there are no reports of any special diamonds, and nothing that could possibly catapult a country with such historically low prices into the premier league. The Kimberley Process seemed not to notice.

I did, however, and although by then I was no longer associated with the Kimberley Process, I was following its work as well as the reports of a UN Panel of Experts. The panel had been appointed by the Security Council to monitor Liberia's compliance with a variety of requirements imposed after the departure of Charles Taylor. When I remarked on the unbelievable figures, the Panel said in its June 2011 report:

> Mr Smillie states that the most troubling issue is the significant increase in value of exports with no increase in the volume and the "incredible increase" in the per carat average value of exports . . . He notes that the average export value per carat was more than double the figure for 2009 of any Kimberley Process member country with the exception of Lesotho. Historically, however, Liberia has been an exporter of low-value goods and the explanation of the Government

> of Liberia for this increase in value "does not sufficiently substantiate the government data." He also states that the increase in export value "suggests that large quantities of low value diamonds are not being legalized at the export stage" . . . The Panel will continue to investigate these issues.[17]

By the end of 2012, the UN Panel had almost nothing good to say about the Liberian diamond sector. Most, if not all, of the ten regional diamond offices had been closed and internal controls were essentially non-existent. The report said that better diamonds were being smuggled out to Sierra Leone, while an estimated 10,000 illegal Sierra Leonean diggers were operating in Liberia. The "good" news was that Liberian per-carat averages had fallen to $403 in 2011 – still, however, second only to Lesotho. If that number continues to ring hollow, so does the figure of 10,000 illicit diggers from Sierra Leone. In 2011 Liberia exported only 40,000 carats in total. If an equivalent amount had been dug up by Sierra Leoneans and smuggled back to their country – representing, in other words, half the country's hypothetical annual production – it would represent only four carats per digger, per year – hardly a great incentive for crossing the border to become an illicit diamond miner.

The UN Panel made a presentation on the dismal state of Liberian diamond affairs to the Kimberley Process in June 2012 and said that "bold thinking" would be required if the problems were to be fixed. The Kimberley Process organized a review team to visit Liberia in 2013, mandated to come up with something akin to bold thinking. For its part, the Liberian government said the problem was a lack of capacity and money, not of the will to implement the plan that had been agreed in 2007.

India

The KP statistical database suggests a variety of other problems in the diamond trade. Dubai, for example, has become a very important diamond trading hub in recent years, usurping some of the role that Antwerp once played and hiring senior Belgian diamantaires to do it. It makes sense, therefore, that Dubai – listed in KP statistics as the United Arab Emirates (UAE) – would import growing volumes of rough diamonds from producing countries such as Russia, South Africa, and the DRC. It also makes sense that there would be a brisk import/export trade with other hubs such as Antwerp and Switzerland. What makes less sense is the importation into the UAE of large volumes of rough diamonds from a country like India. Rough diamonds go to India to be cut and polished. There may be a modicum of trade, and some rough may be returned to a seller via Dubai for perfectly valid reasons. But when India ships $573 million worth of rough diamonds to the UAE, as it did in 2011, something else is going on.[18] And when it imports more rough diamonds in a year than are produced by all of the world's diamond mines, as it did in 2012,[19] something else is definitely going on.

Some of it is undoubtedly what is known in the business as "round-tripping." Round-tripping involves importing and re-exporting the same parcel of diamonds through a company's offshore office in a tax-free haven like Dubai in order to demonstrate inflated turnover, giving a company access to better bank financing. It can also be used for transfer pricing purposes – shifting profits from one tax jurisdiction to another – and other forms of tax evasion and money laundering.

A 2008 KP review of the United Arab Emirates noted a variety of statistical issues and said that "the practice of transfer pricing could possibly be widespread." It added that "this is an issue that deserves wider consideration within the KPCS."[20] The review team was looking at 2007 import figures from India, figures that would grow fourfold over the next four years. Whatever the figures implied, the phenomenon grew by almost 100 percent per annum after 2007. It might have "deserve[d] wider consideration within the KPCS," but it didn't get it.

Where polished goods are concerned, the Indian government put a damper on round-tripping – an illegal activity – at the beginning of 2012 by imposing a 2 percent import duty. As a result, Indian imports of polished diamonds dropped from $20 billion to less than $6 billion in a single year.[21] This did not put a complete end to the practice in the polished sector, but it made an obvious and significant dent. Where rough diamonds are concerned, however, little changed.

As part of a regulatory system, a database is only as good as its contents, and it is only as useful as its regulators are able and willing to act on obvious irregularities. The value in the KP database is its availability for the scrutiny of all members and its capacity for the triangulation of data between and among trading partners. This is a powerful tool, but if tools are not used, they are not much good. A solution was found for the problem in Guinea, but it took four years to get it written down and accepted, and, given the KP's lack of follow-up in other areas, it remains to be seen how effective it will be. It took a UN Expert Panel and other outsiders to point out the deterioration of Liberian internal controls and the fantastical export statistics, and at least three years before the KP could mount a team to address the problem. Solutions may be years in the offing.

The KP's sluggish response to the statistical problems in Guinea and Liberia was a problem of power and politics in the sense that data analysis was being done by a scattered committee that made its decisions and recommendations on the same basis as all other KP decisions – by consensus. Before a problem could be discussed, there had to be agreement that there was, indeed, a problem. Because there was no secretariat, no staff, no support, and no back-up, data analysis was done on a voluntary basis by members of this committee, many of them unfamiliar with statistics and all of them busy with their day jobs. Although there had been calls as far back as 2005 for the creation of a secretariat with the professional skills needed in this area, some governments resisted the idea with bulldog determination. A special working group struggled with the issue between 2010 and 2012, knuckling under to a demand from some governments that any "secretariat" had to be called something else.

At the end of 2012, an agreement was finally reached on what was called an "administrative support mechanism" to be managed entirely by the private sector. The coordinator would be the World Diamond Council, and the functions of the administrative support mechanism would be divided between industry coordinating bodies in Antwerp, India, Ghana, and Israel.

Outsiders can be forgiven for thinking that this looks like a recipe for further confusion, if not conflict of interest. However it works out, it demonstrates a clear unwillingness on the part of some governments to support and equip an effective regulatory system.

Compliance: the definition of a conflict diamond

It was important from the outset for the KPCS to be clear on meanings and definitions in its basic provisions, although, as this chapter has demonstrated, even the clearest of problems seems to generate a multitude of interpretations, especially when it comes to action. So it was with the definition of conflict diamonds. As noted above, the KPCS document defined the term this way:

> Conflict diamonds means rough diamonds used by rebel movements or their allies to finance conflict aimed at undermining legitimate governments, as described in relevant

> United Nations Security Council (UNSC) resolutions insofar as they remain in effect, or in other similar UNSC resolutions which may be adopted in the future, and as understood and recognised in United Nations General Assembly (UNGA) Resolution 55/56, or in other similar UNGA resolutions which may be adopted in future.

The salient point in this definition is the concept of rebel movements undermining legitimate governments, and the wedding of this concept to UN Security Council resolutions. In 2002 this seemed both good enough and clear enough. The purpose of the KPCS was to bring an end to the horrific conflicts in Sierra Leone and Angola. Rebel armies in the Democratic Republic of Congo had not been sanctioned for their diamond-related activities, but it hardly mattered because the KPCS aimed to regulate all diamonds everywhere. So diamonds from the DRC and neighboring countries where conflict and diamonds overlapped borders did not need a special Security Council resolution.[22]

It did not occur to most of those involved in the KPCS negotiations that a member state might use its own armed forces to beat, rape, rob, torture, and kill its own citizens in furtherance of KPCS minimum standards. If it had, the definition of conflict diamonds and the provisions of the KPCS would surely have been different. As it turned out, however, the effort to change the definition after the fact and to include human rights considerations would become one of the most divisive issues in the Kimberley Process.

Angola

The first sign of serious human rights abuse in the diamond fields of a KP member state was not long in coming. Angola's Lunda-Norte Province has a 770-km border with the DRC, and, like so many borders in Africa, it is difficult, if not impossible, to police. The rebel UNITA forces had welcomed diamond diggers from the DRC, and when the civil war ended, more came of their own accord, driven by hunger, poverty, and hopes of a better life. Somewhere between 300,000 and 400,000 illicit alien diggers showed up in Angola between 1990 and 2002. And then the Angolan government decided to get tough.

In an exercise that the military and police called *Operação Brilhante*, an estimated 256,000 illicit diamond diggers were rounded up and expelled by mid-2005. Most were from the DRC, and most were subjected to serious human rights abuse in the process. They were robbed, and many were beaten and subjected to body cavity searches and the use of forced emetics. Reports written by Human Rights Watch, Médecins sans Frontières, Partnership Africa Canada, and others documented cases of brutality, rape, and murder. Tens of thousands were force-marched to the border. The UN Office for the Coordination of Humanitarian Affairs (OCHA) was obliged to set up refugee camps on the DRC side of the frontier, and the African Union's Commission on Human Rights publicly criticized Angola for its handling of the matter.

The Angolan government admitted to excess, and even apologized, but nothing much changed. The Angolan military and police, steeped in violence and coming out of a 25-year civil conflict, would have been difficult to control even if the government *was* serious. They and private security firms were used to shooting first and asking questions later, regardless of the nationality of those digging. So, despite the protests, tens of thousands of expulsions continued through 2008 and 2009, accompanied by violence, torture, theft, and forced marches over hundreds of miles. In 2012, Human Rights Watch issued a 50-page report on the continuing problem entitled, *"If You Come Back We Will Kill You": Sexual Violence and Other Abuses Against Congolese Migrants During Expulsions from Angola*. And while former diggers were being driven in one direction, prospective diggers were crossing unguarded stretches of the border back into Angola, perpetuating the cycle.

A lesson in this is that policing alone, whether it is done well or badly, is not enough to deal with the problem of illicit diamond mining in poor countries. This is not a new lesson. It was as true in colonial times as it is today. It is a lesson that has been learned and forgotten repeatedly in countries where alluvial diamonds are found. There is no obvious or simple solution, but chapter 7 will delve into the development issue and the role it can play in answering this century-old problem.

The issue in Angola is not whether a government has the right to expel illegal aliens or to halt illicit digging. The former is the right of any government, and the latter is a responsibility under the provisions of the KPCS. The issue is one of tactics if they are based on full-blown human rights abuse. NGOs involved in the Kimberley Process said that it made no sense to replace one kind of diamond-related violence with another. They appealed to the industry for support in pressing the Kimberley Process to include the observance of basic human rights as an integral part of certifying diamonds as conflict-free. But nothing happened. A KP review team visited Angola in 2005 but it was not permitted to visit the artisanal mining areas, and it simply noted that Operação Brilhante was taking place. It said nothing about human rights abuse, observing only that "from what the team members could see in the areas they visited in Lunda provinces, indeed there seemed to be much control by the government security forces over the movement of people there."

Much control indeed: "The Team further observed the commendable efforts by the Angolan Government to restructure the diamond industry in the short period following the end of the protracted civil war in 2002, with a view to establishing full control over the sector."[23] A 2009 KP review was even more blank on the subject of human rights.

Zimbabwe

The Zimbabwe diamond story is as complex as it is tragic, but the basic facts are simple enough. Zimbabwe, a small diamond-producing country, was an early member of the Kimberley Process. In June 2006, there was a new diamond strike at Marange on the country's eastern border, and soon the area was flooded with people. The find promised to be huge. Zimbabwe's economy at the time was in tatters. The inflation rate that year was running at 1,200 percent and the Zimbabwe dollar had so little value that it was redenominated by shaving three zeros off the old currency. That didn't help much because two years later it had to be done again, this time making the new dollar the equivalent of $10 billion of the old. By then, inflation was running at a staggering 471 billion percent.

The people flooding into Marange were looking for something with lasting value, but President Robert Mugabe and his ZANU-PF colleagues saw only the theft of a resource that could save Zimbabwe's economy and perhaps the regime. In October 2008 the government sent in the police and military. Subsequent reports from journalists, local civil society organizations, and others such as Human Rights Watch put the death toll at 200. Fleeing diggers had been shot down, some from helicopters. Hundreds were arrested, beaten and tortured.

Zimbabwe appeared to be losing control of the wider diamond scene as well. The bulk of Zimbabwean diamonds are easily recognizable for their distinctive brown color, and KP reports had them showing up for certification as far away as Sierra Leone and Guyana. Arrests were made in Dubai and India of smugglers carrying large hoards of Zimbabwean diamonds. Just across the border in the Mozambican town of Manica, a rush of new diamond buyers – many driving cars with South African license plates – set up shop to evaluate and purchase goods brought in by syndicates of Zimbabwean military officers.

The United States and the European Union placed embargoes on Zimbabwean diamonds. Civil society organizations called for Zimbabwe's expulsion from the KPCS and for the inclusion of a human rights provision. The Council of EU Foreign Ministers noted "with concern the growing trade in illicit diamonds that provide financial support to the regime. In this context, it also condemns the violence inflicted by state sponsored forces on diamond panners and dealers at Marange/Chiadzwa."[24]

The Kimberley Process didn't manage to get a team into Zimbabwe until the end of June 2009, nine months after the massacre. It did corroborate most of what had been reported by the media and NGOs, but before its report could be discussed within the KP, the Chair of the Kimberley Process that year – Bernhard Esau, Namibia's Deputy Minister of Mines and Energy – flew to Harare, claiming to be heading his own fact-finding mission. He held a private meeting with Robert Mugabe and then held a press conference: "Yes, there are members . . . trying to convince other members to suspend Zimbabwe," he told the media, "but we will not entertain such [calls]." He dismissed human rights abuse as beyond the KP's mandate, and he even downplayed the smuggling at Marange.[25] Esau would demonstrate two things during his tenure as Chair. The first was a strong penchant for ill-informed meddling. The second was a determination on the part of his government, and that of South Africa, to protect Robert Mugabe and his murderous regime from all criticism and any restrictions on their new-found diamond wealth.

I realized then what I should have seen months, if not years, earlier: the Kimberley Process would not ever deal efficiently or effectively with this issue or with any other that involved the slightest political or economic discomfort for one of its members. After a decade working on this issue, I left my position as Research Director at Partnership Africa Canada and wrote to Kimberley Process members, telling them that we had let a significant undertaking slip away: "The KP has been confronted by many challenges in the past five years, and it has failed to deal quickly or effectively with most of them . . . In each case the issue has had to become a media debacle before the KP would deal with it (if at all), and in the case of Venezuela, we have effectively condoned diamond smuggling – the very thing we were established to prevent."

> Perhaps worse, we refuse to deal with human rights abuse in alluvial diamond mining, surely a fundamental issue for a body that aims to stop "blood" diamonds. For every hour we spend dealing with issues of pro-forma KP compliance, we devote four hours to argument about why and how to avoid real issues . . . There is a basic truth: when regulators fail to regulate, the systems they were designed to protect collapse. In this case, the diamond industry, which means so much to so many, is being ill-served by what has become a complacent and almost completely ineffectual Kimberley Process. Without a genuine wakeup call and the growth of some serious regulatory teeth, it leaves the industry exposed, vulnerable and perhaps, in the end, unworthy of protection.

The Zimbabwe issue and the debate about human rights dragged on for three more years. A former South African Chair of the KP, Abbey Chikane, was appointed as an external monitor to enforce a list of KP provisions that had nothing to do with human rights, smuggling or the flourishing of new diamond mining companies set up in Marange and controlled by ZANU-PF insiders. On one occasion, a brave Zimbabwean whistle-blower attempted to convince the monitor of continuing human rights abuse. Rather than investigate the allegations, however, KP monitor Chikane turned the whistle-blower in to the Zimbabwean authorities to face beatings and imprisonment. Never once did Chikane mention human rights or corruption, and never once did he query or delay the export of a parcel of Zimbabwean diamonds.

NGOs and others did manage to crank up the debate on human rights, however. One proposal for added wording to the existing definition started with wording from the UN Universal Declaration of Human Rights: "Whereas disregard and contempt for human rights have resulted in barbarous acts which have outraged the conscience of humankind, the Kimberley Process shall promote respect for these rights and shall require their effective recognition and observance in the diamond industries of participating countries and among the peoples, institutions and territories under their jurisdiction."

When the United States took the Chair of the Kimberley Process in 2012, it set for itself a two-item agenda: the creation of a professional secretariat, and a redefinition of the term "conflict diamond." As noted above, the KP did agree that year to the creation of a confused, private-sector-managed "administrative support mechanism." By the end of her tenure as Chair in 2012, however, Ambassador Gillian Milovanovic had reduced her expectations for a revised definition to: "'Conflict diamond' is a rough diamond either used by rebel movements or their allies to finance conflict aimed at undermining legitimate governments, or otherwise directly related to armed conflict or systematic armed violence."

There was a lengthy explanation of what this meant, but it did not include any reference to human rights. The opposition to a human rights provision from some industry members and a range of governments – led by South Africa and India – had been so fierce during her tenure that this was the best she could propose. She took the wording with her to a diamond industry love-in at Victoria Falls in Zimbabwe in October 2012, hoping to get support before the Plenary Meeting in November. Robert Mugabe attended, as did former South African President Thabo Mbeki. Industry representatives were there in droves, eager for a nod from the Zimbabwean authorities if and when the hoped-for lifting of restrictions occured. Even Milovanovic's watered-down definition was rejected. She was the target of insults and told she should resign. And when the Kimberley Process Plenary met the following month in Washington, it "commended" Zimbabwe, lifted all special measures, and, in something akin to putting a viper in a kindergarten, appointed it to the Working Group on Monitoring.

Three months later, in January 2013, as diamonds flowed out of Zimbabwe without restriction, the country's Finance Minister, Tendai Biti, told journalists that the government had nothing left in the bank. "Last week when we paid the civil servants, there was $217 [left] in government coffers," he said. A spokesman for the Movement for Democratic Change explained why: "The diamond wealth is going to ZANU-PF machinery and its war chest."[26]

In June that year, former Zimbabwe Mines Minister Edward Chindori-Chininga sent the media a copy of the report of a Parliamentary Committee investigating the diamond industry. It documented devastating levels of corruption, mismanagement, and violence in the diamond sector over a four-year period, and the disappearance of hundreds of millions of dollars in diamond revenues from government coffers.[27]

The day after he sent the report to the media, Chindori-Chininga was killed in one of the "accidents" so common to enemies of the Mugabe regime. His car, a silver-colored Jeep Cherokee, allegedly failed to navigate a T-junction on the Raffingora–Mvurwi Road and hit a tree. Pictures of the car showed that the windscreen was intact and the airbags had not been deployed.[28] But Chindori-Chininga was dead.

As for the KP debate over the definition of a conflict diamond, the Plenary's communiqué said: "After lengthy discussions, no consensus was reached on whether or not to change the definition,"[29] and the whole discussion was shelved for another year.

Despite the limited definition of conflict diamonds in the basic KPCS document, the original intent was clear enough. The second paragraph in the preamble of the Kimberley Process recognized "the devastating impact of conflicts fueled by the trade in conflict diamonds on the peace, safety and security of people in affected countries and *the systematic and gross human rights violations* that have been perpetrated in such conflicts."[30] The WTO waiver of 2003 used precisely the same terminology as did the KP-related legislation of the United States and other member countries. The wording may have connected the human rights issue with "such conflicts" but, had anything like Angola or Zimbabwe been in sight, there would have been no ambiguity on the need to deal more specifically with gross human rights violations.

At the end of 2011, Global Witness, one of the two NGOs at the forefront of the conflict diamond campaign, announced that it was through. It had been part of the KPCS negotiations and, along with Partnership Africa Canada, had been part of the effort to make the Kimberley Process effective for nine years. Now it was quitting. In a press release, it said:

> The world has moved on but the Kimberley Process remains stuck in time. Ever more insular, the KP has spent the past few years lurching from one shoddy compromise to the next in a manner that strips away its integrity and undermines its earlier achievements. The KP has failed to deal with the trade in conflict diamonds from Côte d'Ivoire, breaches of

> the rules by Venezuela and diamonds fueling corruption and state-sponsored violence in Zimbabwe.
>
> Most recently, the decision to endorse unlimited diamond exports from named companies in the Marange region of Zimbabwe – the scene of mass killings by the national army – has turned an international conflict prevention mechanism into a cynical corporate accreditation scheme . . . The Kimberley Process's refusal to evolve and address the clear links between diamonds, violence and tyranny has rendered it increasingly outdated. It is time for the diamond sector to start complying with international standards on minerals supply chain controls, including independent third party audits and regular public disclosure. Governments must show leadership by putting these standards into law.[31]

One man's meat, of course, is another man's poison. A critical change had taken place over the decade between the KPCS negotiation era and the later debate on human rights. In 2003, NGOs held the moral high ground; by 2013 they could be dismissed by Zimbabwe and its supporters as a nuisance. The KP deficiencies of 2003 had actually become an enabling factor for some of the dominant interests a decade later.

Martin Rapaport, one of the first industry leaders to recognize the conflict diamond challenge (see chapter 3), became wild with anger over the situational ethics at play in the Kimberley Process and some parts of the diamond industry. When the November 2012 KP Plenary postponed yet again any serious consideration of human rights, he produced a powerful attack he called "Business Ethics 101: Moral Clarity and the Diamond Industry."[32]

"Sometimes it's hard to tell right from wrong," he wrote:

> And sometimes it's not . . . Some people don't want us to tell right from wrong. They prefer that the diamond industry avoid ethical issues and concentrate on "business only." Often this is because they and their friends make windfall profits selling questionable products. When serious ethical

> matters such as severe human rights violations get in the way of their profits, they manipulate the truth through misrepresentation, denial, half-truths and confusion to protect their business interests.

He detailed the killing, torture, and human rights abuse in Zimbabwe, tearing strips off the military and the new mining companies with secret ownerships. He called the government "a gang of thieves disguised as politicians stealing diamonds from their own people in one of the poorest countries of the world. Suppliers so evil that the U. S. Government has slapped Office of Foreign Assets Control (OFAC) sanctions on them and you can't legally import their diamonds into the U.S." He heaped scorn on Indian diamond buyers and others lining up to get their hands on Zimbabwean stones. He spoke of the October 2012 diamond conference in Zimbabwe, "designed to legitimize Mugabe, [Mining Minister] Mpofu and their gang of thieves," where

> leaders of the diamond industry fell all over themselves, toadying up to the bad guys. Eli Izhakoff, chairman of the World Diamond Council went so far as to say he would petition the U.S. and EU governments to drop sanctions against the Zimbabweans . . .
>
> [And] what about the Kimberley Process? Isn't the KP supposed to certify that diamonds are free of human rights abuse and violence . . . Here is what Ambassador Gillian Milovanovic, US Chair of the KP had to say at the Zimbabwe conference: "KP certification is not designed to address human rights, financial transparency, economic development or other important issues."

The Rapaport bile flowed like battery acid: "From the KP's perspective, it's perfectly all right for Mpofu and his ZANU-PF gangsters to kill as many people as they like, or steal as much money as they like. The KP is only interested in cases where rebel forces use diamonds to attack a government

. . . The WDC is pouring poison into the well from which we all drink," he said. And he spoke about socially responsible consumerism: "Socially responsible consumerism is not just a trend; it's a long-term lifestyle commitment . . . How long do you think our industry can get away with selling diamonds involved in human rights violations? We must separate ourselves from the bad diamonds. It's not just an ethical or legal issue. It's survival."

CHAPTER SEVEN

Development

During the first two-thirds of the twentieth century, diamonds were mined almost exclusively in colonial territories or in the two countries operating under apartheid – South Africa and South West Africa. The only exceptions were small volumes of diamonds produced in Brazil, Venezuela, and Liberia. One *could* say that diamonds helped to build infrastructure and foster development in the colonial territories, but such a generalization would rest on shaky ground. The Portuguese left behind a 15-year legacy of colonial war, social upheaval, and political wreckage in each of their African colonies. Whatever one might say about French colonialism, in the three or four of their colonies where diamonds were mined, production was limited. In Tanganyika, John Williamson created a beautiful campus around his diamond mine, complete with schools, a hospital, a dairy, and other facilities, but the benefits didn't extend much beyond his own employees and the main gate, and he did it – at least in part – because the alternative was to give the bulk of his profits to the colonial government, something he was loath to do. In Sierra Leone diamonds did contribute to some of the colony's development. Taxes were paid into the common exchequer, and, like its counterparts in other countries, Koidu grew in 25 years from a tiny hamlet into a sprawling town with new jobs, opportunities, and potential. But these were patchy and skewed. When I arrived in Kono District in 1967, Koidu Secondary School – the only one in the country's fourth-largest town – was only two years old, there were no sewers, and if the "highway" to Koidu had ever been paved, there was no evidence of it. Kono District was and remains the least-developed area of the country, a phenomenon common to the diamond areas of other African countries.

Diamonds flowed out of the Belgian Congo like water from an open hydrant, yet by Independence Day the country had produced only a handful of university graduates. And while the Oppenheimer family in South Africa developed a solid reputation as opponents of apartheid, the system remained in place throughout most of the twentieth century. "Apartheid was baked hard into the mining industry, because that's where it originated," says Cyril Ramaphosa, a leading figure in today's South African politics and business.[1]

So whatever claim diamonds may have to being a serious generator of development, it post-dates the independence movement of the 1960s and 1970s. And, as chapter 3 has shown, with very few exceptions – Botswana and Namibia being the most notable – the picture is not a pretty one. It isn't that diamonds contributed little to development in the 30 or 40 years following independence. It's that in most of the major producing countries of Africa they fueled corruption, bad governance, human rights abuse, and the most horrific of wars.

It was not until the end of the diamond wars and the creation of the Kimberley Process Certification Scheme in 2003 that the first and most comprehensive possibility of diamonds as a force for development at last became apparent. This chapter will examine the myths and realities of that potential.

Some numbers: mining revenues and equity positions

Diamonds are mined in about 30 countries. Of these, 27 are members of the Kimberley Process, while a few such as

Gabon remain outside the fold. Of the 27, all but 3 – Russia, Canada, and Australia – are developing countries, and, of those, 17 are in Africa. In 2012, 57.4 percent of the world's diamonds by weight were produced in developing countries, and 59.3 percent by value.[2] The value of diamonds exported from developing countries in 2011 was $10.2 billion, and more than 99 percent of this was from Africa. This is a big number, but it is misleading.

In Africa, the export tax on diamonds is typically 3 percent of the value of the goods. Critics frequently argue that governments are letting a valuable non-renewable resource go for a pittance. Whenever a government raises the tax, however, diamond exports have a not-very-mysterious way of plummeting. It happened in the DRC when government raised the export tax one point over the rate in the Republic of Congo. Diamonds simply moved across the River Congo to the more agreeable tax regime. At the end of 2009 Sierra Leone imposed a 15 percent export tax on stones valued at more than $500,000. This was quickly followed by a huge increase in the per carat average export price from Liberia, described in chapter 6.

Export taxes are not a government's only source of diamond revenue, of course. Prospecting licenses, mining licenses, trading and export licenses, and income tax all add to a government's revenue if it has the wherewithal to administer and collect the fees. Botswana, often held up as the exemplar of best practice where diamonds are concerned, has actually ignored standard development orthodoxy by taking an ownership position in the country's diamond mines. It began with a 15 percent share in Debswana, then raised it to 50 percent in the 1970s. The 50-50 arrangement with De Beers is unique, not least because the Government of Botswana eventually bought 15 percent of De Beers as well. Next, the government persuaded De Beers to move its $6.5 billion sorting operation from London to Botswana, a move completed in 2013. The benefits to Botswana have been enormous in terms of income and job creation. But Botswana is not just atypical of diamond-producing countries, it is unique. In addition to having a government that understands the diamond business, Botswana produces more top-quality diamonds than any country on earth. $5.5 billion worth of diamond production – the 2011 figure – will buy a lot of friends and influence.

Angola, the world's fifth-largest diamond producer by value in 2012 has, like Botswana, a great deal of clout with foreign investors, but the approach has been different. Angolan diamonds are both kimberlitic and alluvial in nature, and although much of the alluvial production is generated by artisanal miners, there are also industrialized alluvial operations. Angola therefore attracts both large and medium-sized international mining firms. Angola's state diamond company, Endiama (the name is derived from Empresa Nacional de Diamantes) acts as a kind of trustee for the country's diamond rights, and requires that all foreign investments be established jointly with local firms. Typically, Endiama will take 51 percent ownership in new kimberlite operations, although, depending on the size and circumstances, this has proven to be negotiable. Alluvial mining operations will typically require that at least 51 percent should be in Angolan hands, with a negotiated split between Endiama and any other Angolan partners.

This appears to work well, up to a point, especially for local "partners." In fact, Angolan companies usually contribute little to a partnership beyond Angolan faces and political cover for the international investor. Most have no mining expertise and few invest money. This has two consequences. The first is that the foreign investor must put up all of the cash in return for less than half of whatever profits are generated. This, in turn, makes the cost of doing business in Angola steep. It ensures that the investor's eye will always be fixed unequivocally on the bottom line, and that frivolities – such as corporate social responsibility projects, for example – will be of ancillary concern. The second consequence is that a handful of Angolans, many connected to the ruling party and the military, have become exceedingly wealthy by creating companies and joint ventures with foreign investors. For a very few Angolans, diamonds have become a get-rich-quick opportunity, with almost none of the risks found in common or garden variety mining and investing.

How well does this work for the average Angolan? The answer is not clear, but there are indicators. In 2012 Angola ranked 157th out of 174 countries on the Transparency International corruption perception index, a slight improvement over 2011 when it ranked 168th.[3] It ranked 148th out of 179 countries on the United Nations Human Development Index in 2011, but this rating is skewed somewhat by the country's relatively high gross national income per capita, generated by oil and diamonds. An indication of how far down the mineral wealth trickles can be seen in Angola's life expectancy statistics. An Angolan born today can expect to live slightly more than 51 years. Only four countries on earth have a lower life expectancy.[4]

Neither Botswana's nor Angola's style of beneficiation is open to many African countries, especially those where artisanal mining represents a large proportion of production, and where major companies like De Beers are notable for their absence. In the past, many governments did take an ownership position or even nationalized a mining operation, almost always with poor results. Tanzania nationalized De Beers' Williamson Mine in 1971 but, after a disastrous 24-year run, invited the company back, holding on to only 25 percent of the equity for itself. In 2006, Sierra Leone's Minister of Mineral Resources spoke about nationalization:

> We made that mistake with the Sierra Leone Selection Trust (SLST) when we nationalized it and called it the National Diamond Mining Company (NDMC). We took 51% of the shares of the company, and were then forced to pay 70% tax to the government. But the government still wanted a share from the company's finds – half of the remaining 30%. The government did not have money to pay for its shares, so they were paying from dividends they got from NDMC each year. It was absurd. NDMC was forced to pay dividends each year, and so there were no reserves, no savings, no investment by the company. I was an official of the NDMC. We were not able to replace old machines; there was no money to invest in equipment. Everything collapsed in short order. That is the story of one failed attempt at nationalisation. It doesn't work; we should not even talk about it.[5]

Artisanal miners

Approximately 16 percent of the world's diamonds are produced by artisanal miners. An artisanal miner is a person – usually a man but often a child – who digs diamonds by hand. The work is hard, dirty, and dangerous. It involves digging for alluvial diamonds that have been scattered across hundreds of square miles over 1,000 millennia. Sometimes they are close to the surface, but they can also be 20 or 30 meters underground. With rudimentary tools and pumps and little knowledge of safety practices, the work is dangerous. Some artisanal mining is done in fast-running rivers that are even more hazardous. And because gravel must be washed in order to find the gems, much of a digger's day is spent standing in stagnant water.

Artisanal diamond mining is environmentally unsound. Huge swathes of arable land are rendered useless because the topsoil is shovelled away leaving innumerable craters and fetid ponds in their wake. Rivers are crudely dammed and diverted, destroying fisheries and polluting water. It is said that the only man-made object visible from space is the Great Wall of China, but artisanal diamond fields rival that claim. If, using Google Earth, you start at "Kenema Airport Sierra Leone" and then go to "Koidu Sefadu Sierra Leone" you will see vast stretches of countryside laid waste by artisanal mining. In the opening Koidu view, bottom center, you will see the country's only kimberlite mine, and if you zoom in and move around the town-site you will see countless ponds and craters and stretches of ground where nothing grows. These are the artisanal diamond pits – past, present, and future.

Most artisanal diamond diggers work in the informal sector. "Informal sector" is a polite way of describing illegal behavior. Diggers operate illegally – in the informal sector – for a variety of reasons. One may be that artisanal mining has been outlawed, as it was for many years in Angola. Another may be that diggers are working as illegal aliens in another country, as has been the case for decades with tens of thousands of Congolese in Angola. It may be that diggers are encroaching on the lease of a mining company. Or it may be that even where artisanal mining is legal, diggers cannot afford to buy the required license.

The places where illicit miners gather are notoriously violent, for obvious reasons. People are drawn together from many places and backgrounds. Although few make much money, they deal in a high-value commodity and ever-present security risks. In addition to violence, health conditions are bad and mine sites are incubators for disease. The diggers are prey not only to each other, but to economic vultures who know better than they what their finds are worth. And of course the artisanal diamond fields of Sierra Leone, Angola, the DRC, and Côte d'Ivoire are where conflict diamonds were spawned, for the obvious reason that they could be mined without great difficulty or expense.

Historically, governments have dealt with the problem in a variety of ways, all of them regulatory, and, as previous chapters have shown, many of them extreme in their prejudicial nature. In the Sierra Leone I knew in the 1960s, illicit diggers were rounded up by the army, trucked 100 miles away, and dumped in the forest. Most soon returned. In Angola, illegal aliens who are found mining diamonds are treated more harshly: they are beaten, robbed, and force-marched to the border. Many of them soon return as well. Shooting incidents are not uncommon around company mine sites, and in Zimbabwe the armed forces resorted to wholesale slaughter in order to clear diggers away from Marange in 2006.

The Kimberley Process Certification Scheme too is a regulatory instrument. Regardless of its efficacy, it is based solely on the tracking, verification, and certification of diamonds. This is undoubtedly why the negotiations involved in creating the KPCS moved as quickly as they did. Those of us involved in the NGO coalition during those days were often asked by newly joining NGOs whether there were environmental issues in diamond mining. Or labor issues, human rights issues, or child labor. I always answered that these problems undoubtedly existed, but the wars were the most important issue, and if we couldn't deal with the violence, there would be little point in tackling anything else. Unlike many other certification efforts, therefore, the KPCS negotiations were focussed on one thing only and were unencumbered by other issues. The KPCS was essentially a Western approach advocated by Western governments, NGOs, and industry actors, and sold to African and Asian participants as a scheme to protect their interests going forward.

But the development *problématique* was always there. More and better regulation would not change the plight of the artisanal digger in the places where conflict diamonds began. That was and remains a development problem, one that requires development solutions.

Once the KPCS was up and running, Partnership Africa Canada and Global Witness produced a joint report on artisanal mining. It was called *Rich Man, Poor Man* and it found that the million or more artisanal diamond diggers in Africa live in a casino economy where most hope to strike it rich but seldom do. Most live on a dollar a day, which places them squarely in the World Bank's category of absolute poverty. The report saw opportunities for something different, and said that "Real change could reduce the chaos and instability that the diamond fields spawn. At a minimum, diamonds could be the generator of decent incomes for hundreds of thousands of families rather than the centre of unsafe, unhealthy, badly-paid piecework."[6]

The report spoke of creating a wider group of stakeholders to discuss the development challenge: the Kimberley Process of course, but also development organizations like the World Bank, the United Nations Development Program, the African Development Bank, and a broad spectrum of international and African development NGOs. When the report was presented at the Kimberley Process Plenary in 2004, two governments with long and difficult experience of artisanal mining – Sierra Leone and Guinea – said they thought it was a good idea. Others, however, said that, while the issue was undoubtedly important, it had no place in the Kimberley Process. It would dilute the purpose of the KPCS, introducing an esoteric subject of little interest to member states like Belarus, Armenia, or Ukraine.

The Diamond Development Initiative

The subject, however, did not seem esoteric to De Beers, which asked Global Witness and Partnership Africa Canada if they would be willing to co-host, with De Beers, a meeting of interested parties to discuss the matter in greater depth. This was as large a step for the two NGOs as it was for De Beers. Although industry and civil society had found themselves more often than not on the same page during the KP negotiations, this idea – actively working together – was an entirely new concept.

For industry, caught completely unprepared by the conflict diamond campaign, here was an issue where early engagement made sense. In the 1990s, the diamond industry's concept of corporate social responsibility (CSR) was limited to schools and clinics in a company's immediate vicinity, not unlike John Williamson's efforts during the 1940s in Tanzania. By 2005, however, industry leaders were moving beyond simplistic notions of CSR to a consideration of issues more organic to their operations. The Responsible Jewellery Council, described in the final chapter, is an example of this. Engagement on artisanal mining, even if no companies worthy of the description were directly involved, was another.

The first meeting was held in London in January 2005. It was well attended by industry, interested governments, and NGOs, and a further meeting was held six months later in Ghana. There the outline of a new organization was chalked in. The Diamond Development Initiative (DDI) would parallel the Kimberley Process and would in some respects have the same sort of participation. The idea was to bring interested governments, civil society, and industry together to tackle the issue of artisanal mining – not through small one-off projects, but at higher levels of policy and on-the-ground efforts that would lend themselves to replicability. DDI would be an NGO, a charitable organization that could solicit funding from governmental donors, industry, and the public at large.

What follows is a description of three major DDI efforts in its first five years, with a caveat – what authors sometimes call "full disclosure." I was present at the founding meetings of DDI and have served as its Board Chair from the outset, so my description is bound to contain subjective elements and you may want to go elsewhere for additional views.

Structurally, DDI is a non-profit, charitable organization incorporated in both the United States and Canada. It has an international Board of Directors drawn from Europe, North America, Africa, and Australia, representing various parts of the industry, civil society, and academia. It is not an industry creation, but it was agreed from the outset that industry had to be involved, not just in the sense of assuming some responsibility, but in contributing in a professional way to the challenges that DDI would face. Funding would be sought from industry, but care would be taken to balance industry contributions with more traditional sources of development finance. De Beers helped to pay for the creation of a DDI website, for example, and the Tiffany and Co. Foundation helped with important core start-up funding, but more than half of the initial start-up capital was provided by the government of Sweden. In order to ensure distance, corporate donations are largely un-earmarked and are unrelated to any company activity, whether in a country where diamonds are mined artisanally, or elsewhere.

The bulk of project funding during DDI's first five years came from governments that recognized the importance of the challenge: Britain, Canada, Germany, Belgium, and the United States. Remarkably, DDI has also successfully solicited financial support from the governments of African countries where artisanal mining has been so problematic. This funding is also un-earmarked and is not even conditional on DDI programming in the donor country. Speaking from a four-decade perspective in the field of international development, I can say that these contributions from African governments are unprecedented. I am not aware of an African government making an un-earmarked cash contribution to an international development NGO, *ever*.

Human Rights

DDI hired as its first Executive Director Dorothée Gizenga, a Canadian of Congolese origin who had been involved in the Kimberley Process negotiations. In addition to her knowledge of the diamond industry and its key stakeholders, she brought experience of both government and civil society, along with a no-nonsense approach to getting things done. In addition to her mother tongue, Lingala, Gizenga speaks English, French, Portuguese, and Russian, which means that she can talk directly and without translators to almost everyone with a stake in the diamond industry, from *garimpeiros* in Angola and Brazil to the executives of Alrosa in Moscow and governments across Africa. Managing projects is Ngomesia Mayer-Kechom, a Duke University alumnus and a law graduate from Cameroon who found work in a Houston law office too tame.

Hailing originally from Africa, neither Gizenga nor Mayer-Kechom is cut from traditional development-NGO cloth. In developing and managing projects, they do something that many, if not most, senior staff in international NGOs do not. They stay in regular touch with project partners and anyone else with a stake in what DDI is doing. Much is made these days of African cell-phone connectivity. DDI puts it to good use with weekly and even daily calls to stakeholders and to those on whom stakeholders depend. This translates into regular discussions with senior government officials in several African countries, and the development of relationships and a level of confidence that I, frankly, have not seen in other organizations.

DDI has been given Special Observer status in the Kimberley Process, where it works with the 15 member countries in the

Working Group on Artisanal Alluvial Diamond Mining. The governments in question all recognize the challenges they face, but because the KP has no development mandate or budget, DDI has become the most obvious, de facto partner for them.

Chapter 6 described two egregious human rights problems in the diamond areas of Angola and Zimbabwe, and the problem the Kimberley Process has had in getting to grips with the issue. In truth, human rights violations of one sort or another are common occurrences in most countries where minerals are mined artisanally. The abuse occurs most frequently when police or company security forces try to protect a commercial mining operation. But if law enforcement agencies are the perpetrators, artisanal miners are often the instigators.

In formal KP discussions about human rights, the greatest pushback came from African governments. Behind the scenes, however, many recognize and accept that a serious problem exists. What they don't want is to take on responsibilities that they cannot manage, responsibilities that – unfulfilled – will only damage their reputations further. Shrill attacks from human rights organizations seemed only to make matters worse. This is something, however, that DDI was able to discuss with government officials in ways that others could not. The result is an experimental human rights education program, written by Africans for both artisanal miners and law enforcement agencies. The Department of Theatre Arts at the University of Malawi is converting some of the curriculum into a play and a video that can be adapted for a range of audiences and languages in other countries. And interestingly, the governments of countries where human rights abuse has been rampant are among the first to offer themselves for the experiment. If, in due course, the implied threat of more embarrassment can be reduced, this could be a back-door route to the adoption of a human rights standard in the Kimberley Process.

Registration

Estimates of the number of artisanal miners in the Democratic Republic of Congo range between 700,000 and a million. The number is somewhat seasonal and it fluctuates with the general state of the economy, diamond prices, new finds, and the movement of rebel armies. Although it is illegal to dig for diamonds in the DRC without a license, few licenses are sold from one year to the next, which means that all of these diggers, who produce almost all of the DRC's $200-400 million worth of diamonds in a year and are the country's biggest export earners, operate in the black economy. They are illicit diggers. There are several implications. First, they operate in a shadow economy, always vulnerable to predators and the unpredictability of the law. They must sell whatever they find quickly, and at whatever price is on offer. The government collects no fees or taxes from them, and as a consequence of their invisibility, it has no idea where diamonds presented for export at Kinshasa are actually mined. It doesn't even know for certain whether they were mined in the DRC, one of the most basic pieces of information required by the Kimberley Process.

Formalizing the informal sector – bringing illicit diamond diggers into the legitimate diamond economy – is a recurring dream in all countries where diamonds are produced in this way, but it isn't likely to happen by issuing edicts and drafting more regulations. And 75 years of history demonstrate that it will not be accomplished with round-ups, arrests, and human rights abuse.

DDI examined the registration system used in Guyana and believed that it could be adapted to Africa. A manual on the Guyanese system was produced and translated into

French and Portuguese for a 2009 meeting held by DDI in Johannesburg. Representatives of 14 African governments attended, as did an official from Guyana's Geology and Mines Commission who explained in further detail how their system works. It has two primary and commonsensical attributes: first, the registration fee is low; and second, officials go to the miners to register them. In most African countries, it is the reverse: the fees are high and miners must travel great distances to a government office where they might wait a day or two before getting anyone's attention.

DDI offered to run a pilot project in Africa. Three governments volunteered and, not without trepidation, DDI selected the DRC for the experiment. The first step was to have the registration fee of $25 reduced to $5, no small task and one that required the agreement of three ministries. The most compelling argument might have been that $25 × 0 = 0, while $5 × something = something. For diggers earning a dollar or two a day, $25 was simply beyond reach, so the reduction was critical. The next step was taking the registration process to the diggers. Working with Congolese NGOs, small diamond-buying companies, and government, DDI worked out a plan that started with a training program. Then 18 teams of five set off for the diamond areas, sometimes a day's travel or more from the nearest government office. The teams were mixed: two NGO people, two from government, and one from business. This ensured both probity and full transparency. DDI took GPS locators to each of the mining sites so they could be properly logged, and each registrant answered questions about his dependants, length of time in diamond mining, other types of work and income, and additional questions that might be helpful in better understanding the sector as a whole. All of this was logged in registration books and later entered onto computers and downloaded into a databank on a dedicated project website.

The project aimed to register 10,000 diggers in something under 12 months. Given the history of the DRC, the long-standing illicit nature of the entire diamond economy, and the sheer logistical challenges, this was ambitious. When the teams set out on their first trips in mid-2011, however, neither they nor DDI anticipated the uptake. By March 2012, over 100,000 diggers had been registered, a phenomenal achievement and one that put to rest the decades-old canards about how it could never be done, how the artisanal sector could never be reformed or brought into the formal sector.

Quite apart from the low fee and officials actually going to the client, there were other reasons why a digger might agree to sign up. By registering, he now has status – legal status. This means that he has legitimacy and cannot easily be driven off his claim by the military or police. Registration is no guarantee of this, but it has enough promise to make it attractive. And second, a digger no longer has to get rid of a diamond the moment he finds it. Being legitimate now, he can take a bit more time to shop around for a better price. And there are other benefits. The Congolese government obtained information on hundreds of mine sites it knew nothing about, and the project produced a database that can be used for future research, and more importantly for future development projects.

This project turns all received wisdom about the nature of illicit miners on its head. It offers lessons not just to other countries where diamonds are mined artisanally – it offers lessons for other minerals too. DDI's follow-on effort in the DRC, in fact, was to apply lessons from this project to artisanal gold miners in Kisangani to determine which aspects might actually be transferable to a new and larger project it is undertaking with World Bank funding.

Development Diamonds

One of DDI's first projects was an effort to create what it calls "development diamonds." The idea is not unlike the Fairtrade concept, but it aims for broad uptake and replicability that will not depend on babysitting by external "helpers." The idea is simple: create standards of good practice and fair prices for artisanal mining operations; give diamonds from those sites a "seal of approval"; and track those diamonds through a chain of participating traders, cutters, and polishers to the retail trade where consumers can be assured of an ethical product.

The concept is simple, but in practice it is quite the opposite. DDI wanted to create a system that could in due course operate independently of the kind of hand-holding required in a pilot operation. This meant, among other things, that DDI could not, and would not, handle any diamonds itself. It would not engage in any commercial activities. It meant that standards had to be created, which required a great deal of consultation with communities and diggers themselves in an effort to determine what was possible and feasible, what minimum standards could be applied, and how the bar might be raised over time. It meant finding willing partners along the supply chain, and interested retailers willing to take part in the first experiments. It also meant the creation of independent third-party monitoring systems that could do two things. The first is to certify that mining sites meet the standards that have been established, and the second is to track the diamonds after they leave the site, in order to ensure consumer confidence in the final retail product. The complexity in all of this is enormous, requiring great care and sensitivity on the part of DDI, and the involvement of diggers, communities, local government, traders, retailers, and donors, all of whom had – for the first efforts – to suspend their cynicism in order to participate in an experiment that could go wrong at any stage of the process.

DDI has studied the feasibility of a similar effort with small Brazilian diamond mining cooperatives. More experiments will be needed and all of it will have to be carefully documented, evaluated, and costed before there can be claims of success. But given the history of artisanal diamond mining and the little it has so far brought to diggers and their communities, this is an experiment with enormous development potential, and one that will provide important lessons for the future.

Cutting and polishing

A major problem faced by many developing countries is that their raw materials are shipped, trucked, and piped away to places where jobs are created for other people and other countries. This is true of most minerals found in Africa and it is also true of many of the continent's natural resources.

Huge volumes of timber, for example, leave Africa every year as logs. Of local value added, where it occurs, 90 percent is of a primary nature: in sawn wood, plywood, and veneer. These represent a tiny proportion of timber exports. Secondary and tertiary production – furniture, flooring, mouldings – is almost non-existent. There are many reasons for this, some relating to local demand, the domestic investment climate, and international trade barriers. As with textiles, petroleum, and other products, progressive African governments have sought to change the equation by improving the local business climate, investing in infrastructure and education, and making sure that credit facilities are available for local investors.

It is not surprising that diamond-producing countries would want to apply the same principles, encouraging the cutting and polishing of diamonds locally, and even the manufacture of jewelry for export. It seems logical, and it was the logic applied to Canadian diamonds when they were first discovered in the Northwest Territories in the 1990s. The territorial government insisted that the first mining companies on the scene – BHP-Billiton, Diavik and later De Beers – make locally mined diamonds available to the cutting and polishing industry it hoped would germinate. They did, and it did. The mining companies agreed that up to 10 percent of what was mined locally would be made available to local firms. They also funded training programs at Aurora College in Yellowknife in mining, crystallography, diamond grading and polishing, jewelry making, and other related studies.

Despite the incentives, it didn't work. One after another, companies set up shop, stressing the "conflict free" nature of Canadian diamonds, and one after another they went into receivership and bankruptcy.[7] There were several problems, labor perhaps the most serious. Canadian cutters and polishers could only compete against India's much lower labor costs if they were producing greater value. And this could only be achieved if they were cutting bigger or better stones. But the 10 percent arrangement was not based on Canadian companies getting bigger and better stones; they were only entitled to a run-of-mine average: some good diamonds, some not so good. If they wanted bigger and better beyond their 10 percent allocation, they would have to get in line like everyone else.

De Beers Chairman Nicky Oppenheimer put his finger on an inconvenient truth in 2000. "The South Africans have a viable cutting industry," he said:

> but it has taken them 70 years to establish. Look at what it cost the South African government over those 70 years – look at the subsidy the government had to invest to create 1,500 jobs in South Africa. The return will be a complete disaster. Show me a single producing country other than South Africa which has a viable cutting industry attached to it.

> Maybe Russia, but even here I wonder if they have accounted properly.[8]

His point was that, if the government of Canada's Northwest Territories wanted to take advantage of diamond wealth to create new jobs, it should focus on investment opportunities where there is genuine local advantage, not diamonds. What's more, the life expectancy of the Canadian mines is in the 25-year range, raising additional questions about the long-term viability of any cutting and polishing in the region.

Despite Oppenheimer's warning, many of the major players, including Lev Leviev and Oppenheimer's own company, De Beers, bowed to pressure from governments in Southern Africa, following Canada's lead in supporting what is known as local "beneficiation." New cutting and polishing operations were established in South Africa, Namibia, Botswana, and Angola in the 1990s and 2000s. In time, some of these enterprises may succeed on their own, but the collapse of WakeGem, a South African company trading under the name African Romance, provides a cautionary tale. In the mid-2000s, WakeGem borrowed R97 million (US$10.6 m.) from the government's Industrial Development Corporation and another R50 million (US$5.5 m.) from a provincial government agency in order to create jobs in the polishing and jewelry manufacturing sectors. In 2013 WakeGem went into liquidation. The company's Chairman said that the recession of 2008 had been problematic, but he added that beneficiation companies like his "will only succeed if there is a pipeline of raw materials that is guaranteed, and if the funding is able to support a long-term time horizon."[9] The raw material issue was similar to the Canadian problem: it was not only about supply, it was about supply of better stones that might make South African labor competitive with India and China. Four months before the collapse, a member of the Gauteng state legislature echoed what Oppenheimer said years before. This was a "vanity project," he said, "without any prospect of viability without continued subsidisation, creating very few jobs in the process."[10]

Impact benefit agreements

Many of Canada's mining, oil, gas, and pipeline operations are in the north and are on, or run through aboriginal lands. Extractive industries have been the subject of controversy and confrontation for generations, but when diamonds were discovered, something different happened. The territorial and federal governments persuaded the first diamond mining company, BHP-Billiton, to negotiate what they called "impact benefit agreements" (IBAs) with local aboriginal associations and communities. An IBA goes beyond the standard royalty and taxation agreement with government and aims to ensure that additional benefits are delivered directly to affected communities. Agreements typically include royalty-type payments, training opportunities, environmental and cultural protection measures, and job-creation opportunities. The latter may cover hiring quotas for aboriginal workers on and around a mine site, and they may also unbundle contracts for transportation, catering, and other services to make them more accessible to small, local companies. The first IBAs were concluded between BHP's Ekati Mine and five aboriginal associations in 1996 and they set the pace for further agreements with Rio Tinto, De Beers and other companies and minerals.

Some argue that the first IBAs could have been handled differently and that aboriginal associations could have done better. However, there is as much research showing that something new and positive took place. In addition to whatever concrete benefits the IBAs conferred, they did two other things. They gave a new kind of authority and independence to the communities involved, and they gave companies a more sustainable "social license" to operate. Canada's 20-year experience with impact benefit agreements has not yet been picked up in Africa or South America, but there is no reason why it could not be.

Development in poor countries is a complex business. If ending poverty were easy, it would have been done a long time ago. The challenge, however, has been especially difficult in the artisanal diamond fields of Africa, where, despite the great wealth they produce, people are even poorer than elsewhere, infrastructure is more stunted, and social breakdown more common. The big development agencies have tended to steer clear of mining, especially artisanal diamond mining, perhaps because of an impression that there is enough money for the sector to take care of itself. If there is any message in this book, it is that positive change is possible, but it will not happen without clear thinking, solid investment, a lot of common sense, dedication, and hard work. You don't have to dig very far into an issue of *Vanity Fair, Vogue,* or *Harper's Bazaar* to find the diamond ads for Bulgari, Tiffany, Swarovski, and others. The words – *extraordinary, precious, uncommonly beautiful, eternal* – can perhaps be read as a challenge, not just to the industry, but to the governments of countries where diamonds are mined, to organizations whose mandate is development, and to consumers as well: a challenge to strive at the headwaters of the diamond pipeline towards change and standards that are also extraordinary, precious, uncommonly beautiful, and, if not eternal, at least sustainable, ethical, and developmentally sound.

CHAPTER EIGHT

Loose Ends

The wars and Kimberley

In January 2002, the government of Sierra Leone declared that the war in that country was officially over. For all intents and purposes, the war had ended 18 months earlier with the arrest of Revolutionary United Front leader Foday Sankoh. Sankoh had signed a peace deal with the government but he failed to honor it. Pressed by British troops and UN peacekeepers, his RUF collapsed in a hail of the deceit and bullets that had marked so much of his career. Charged with war crimes by the UN-backed Special Court for Sierra Leone, he descended into madness and died in prison from the effects of a stroke before he could be tried.

His mentor, Liberian warlord and President Charles Taylor, was forced from office in August 2003 and was eventually arraigned before the judges of the Special Court in April 2006, charged with 11 counts of war crimes and crimes against humanity. Because of the danger he posed in Freetown, the trial was moved to the chambers of the International Criminal Court in The Hague, where I had the honor of being the first witness when the proceedings began in January 2008. My testimony – about diamonds – was based on what I had learned about the trade, my time on a UN Security Council Panel of Experts, and meetings with Taylor and some of his henchmen in 2000 when he was President. It was a lengthy trial, criticized for its cost and for the courtroom histrionics of

Taylor's defense team. But it was an important trial, because Taylor was the first former head of state to be charged with war crimes since Admiral Karl Doenitz – head of the German government for four days in May 1945 – sat in the dock at Nuremberg.

In April 2012, Taylor was found guilty on all 11 counts – of aiding and abetting murder, acts of terrorism, rape, sexual slavery, pillage, and more. He denied everything and in January 2013 an Appeals Chamber listened to the case his lawyers made for a dismissal of his 50-year prison sentence. A case was made by the prosecution to have it increased.

In September 2013, the Appeals Chamber upheld the original verdict and the 65-year old Charles Taylor was bundled off to Britain, the only country that had agreed to host him in this eventuality – probably for the rest of his life.

In Angola, UNITA Leader Jonas Savimbi was killed in a military ambush in February 2002, and within six weeks the war in that country was over. Given that the two worst diamond wars ended before the inauguration of the Kimberley Process Certification Scheme, it is fair to ask whether the KPCS actually played a role in ending the phenomenon of conflict diamonds. Despite the fact that the Kimberley Process was never able to trace or interdict the conflict diamonds that continued to flow out of Côte d'Ivoire between 2005 and 2011, there is reason to credit it with at least some success. The very bright light that was turned on the diamond industry by the KP negotiations that began in 2000 alerted a dozy industry to its responsibility for some of the crises. De Beers shed all of its "outside" buying offices, the Belgian government tightened its regulations, and diamond buyers addicted to false invoicing and tax evasion altered their behavior. While the KP did not end the wars in Angola and Sierra Leone, Savimbi, Taylor, and Sankoh found it increasingly difficult to move diamonds and to buy weapons. And militarily speaking, because of that, all of them were in a severely weakened state when they met their end.

A decade after the conclusion of these wars, some ask whether a KPCS is still required. And others who criticize its ineffectiveness ask whether the time has not come to wind it up. There are several ways of answering the question. The first is to examine the cost of UN peacekeeping in three countries that were plagued by conflict diamonds. Between 1 July 2012 and 30 June 2013, the combined 12-month budget for UNMIL, UNOCI and MONUSCO in Liberia, Côte d'Ivoire, and the DRC respectively, was $2.4 billion. $2.4 *billion*. A little over half of that was spent in the Democratic Republic of Congo, where it was clearly inadequate in dealing with the ongoing security problems posed by rebel armies. The total UN peacekeeping cost in those three countries and Sierra Leone over the decade between 2001 and 2011 ran to more than $20 billion. If there is to be sustainable peace in diamond mining areas, and if the UN and its member states – each of which contributes to these costs through assessed funding – are to avoid prolonged and recurring peacekeeping nightmares, they must solve the fundamental problems described in this book. An effective regulatory system for rough diamonds is one tool. Development is another.

There is a further consideration. Diamonds have demonstrated their attractiveness to gun-runners and – perhaps more importantly where the future is concerned – to money launderers. Money laundering is an integral part of most criminal activity, including the illegal drug trade, but it is also critical to the successful financing of terrorist activity. Since the 9/11 attacks in the US, the March 2004 Madrid train bombing, and the London bombings of July 2005, Western governments have become much more focussed on this issue. Today, under the terms of the international Financial Action Task Force, financial institutions must carry out specified measures of due diligence for transactions of $15,000/€15,000 or more.[1] It is highly unlikely that governments concerned about a $15,000 bank transaction are likely to want the diamond industry to return to the multimillion-dollar scams of the 1990s. If the Kimberley Process continues to prove itself unable or unwilling to crack down on the smugglers it was designed to stop, some other body is likely to step in.

The Responsible Jewellery Council

The Responsible Jewellery Council (RJC) is not that body, although it does address issues that Kimberley could and perhaps should have tackled. The RJC was created in 2005 by 14 of the larger companies involved in the financing, mining, processing, and retailing of gold, diamonds, and platinum. Its purpose was to develop social, environmental, labor, and human rights standards for its members and to create credible supply chains from mine to market.

The extractive sector and related industries are overflowing these days with guidelines, principles, and standards, but most are voluntary and few have teeth. The RJC is different. Its standards are detailed and cover a broad range of issues from bribery and money laundering to child labor, community engagement, use of hazardous substances, and human rights. What is of interest is that in order to become a member of the RJC, a company must agree to abide by the standards and to undergo an independent third-party audit of its compliance. In a world of voluntary codes, this is pretty close to unique, and in relation to diamonds, it answers some of the more pressing concerns about social and human rights issues.

The weak link in the chain, however, is the supply chain itself. Rio Tinto Diamonds and Bulgari are both members of the RJC. Both have been audited, and both meet the standards. Neither company, for example, uses child labor. But Bulgari is under no obligation to buy solely from other RJC members like Rio Tinto, and even if it does, without some kind of auditable supply chain, a consumer has no guarantee that Bulgari suppliers meet the standards. In other words, Bulgari has no child workers but some of its suppliers might. This was partially corrected in 2012 when the RJC initiated a chain of custody standard for gold and platinum, although as this book was being completed, a chain of custody system for diamonds remained elusive.

By the middle of 2013, the RJC had expanded to 440 members, but these are the largest firms in the business, and even with a chain of custody for diamonds, the RJC could not reach down to the alluvial diamond fields where few of its members operate – where indeed, few companies of any sort exist. This is where the Diamond Development Initiative and its system for "Development Diamonds" will be of importance. DDI and RJC standards are complementary, as are standards for artisanally produced gold developed by the Association for Responsible Mining. The challenges, however, are enormous, and the glass – where artisanally mined diamonds are concerned – is not yet half-full.

Synthetic diamonds

There is an argument that synthetic diamonds could in time change the nature of the industry. The technology to produce synthetic or laboratory-created diamonds has developed rapidly in recent years. The earliest products were intended for industrial use, but good gem-quality stones of 1 and even 2 carats are now being manufactured by a number of companies on a commercial scale. Much frowned-upon by the traditional diamond industry, synthetics nevertheless have advantages. The first is that they are cheaper than natural diamonds – by as much as 25 percent. The second is that they are virtually indistinguishable from their natural cousins. Most diamantaires cannot tell the difference and need special equipment to detect the trace impurities that identify them. Their third advantage is that they come with none of the baggage of natural diamonds: no mountains of earth and rock are moved to obtain them; there is no environmental damage; they contribute to no conflict; there is no child labor involved; and no digger must stand up to his waist in mud for a week to find one of them.

It is estimated that, today, synthetics account for as much as 5 percent of global polished demand. Technologies are changing and improving, and the cost of production is falling. The jury is out on the future of synthetic diamonds, but when 600 of them appeared in 2012 at the International Gemological Institute offices in Antwerp and Mumbai masquerading as natural, warning bells went off throughout the industry. De Beers, which is a mining company after all, and has a stake in natural diamonds, put out an alert: "Trading in misrepresented or undisclosed products, whether inadvertently or not, could cause irreparable damage to reputation [and] undermine the integrity of the diamond supply chain, damaging both trade and consumer confidence."[2]

The amethyst market provides a cautionary tale. Until the arrival of synthetic Japanese and Russian goods in the 1970s, high-quality rough amethysts fetched $40 per carat. Synthetic amethysts flooded the market at less than $5 per carat, however, killing off not only the amethyst mining industry but the market as well. The amethyst, once a valued semi-precious gemstone, today holds little interest for the jewelry-buying public.

Mike Roman, a senior industry executive, once said, "Don't worry about diamonds. The industry will be OK as long as there are three things: Christmas, sex, and guilt."[3] That may well be true. The industry has survived world wars, apartheid, the blood diamond campaign, and even the downsizing of De Beers. Perhaps, if it must, the Kimberley Process will do what it was designed to do, or new forms of regulation will come into play. Organizations like the Responsible Jewellery Council and the Diamond Development Initiative will help to solve other problems. Matt Runci, former CEO of Jewelers of America, speaks of another factor, one almost never mentioned in the Kimberley Process meetings where consideration of human rights was rejected: the consumer. Authenticity and integrity, Runci said, are essential in today's commercial world: "It is a concern in almost every industry – coffee, tea, timber, textiles and any extractive industry. It's actually harder to identify industries where these issues are not a concern, than where they are." Earning and protecting the trust of consumers is no longer about pricing and advertising. Today, he said, jewelers must be concerned about social, ethical, and environmental responsibility as well: "Instead of asking 'Why don't you pick on oil?' we need to address the criticism and lead the change ourselves." If the industry waits until consumers are asking more actively about social issues, he said, "It will be too late."[4]

Notes

INTRODUCTION

1 As of mid-2013, there were 54 member states, plus the 27 countries covered by the European Union.

1 THE GEOLOGY AND HISTORY OF DIAMONDS

1 Pliny the Elder, *The Natural History*, Book 37, ch. 15, translation by John Bostock and H.T. Riley (1855): www.perseus.tufts.edu/hopper/text?doc=Perseus%3Atext%3A1999.02.0137%3Abook%3D37%3Achapter%3D15, accessed Oct. 24, 2012.
2 An item of status and plunder for much of its known 800-year history, the Koh-i-noor became, briefly, a focus of controversy in February 2013 when British Prime Minister David Cameron visited India and was asked to return the diamond. "I don't think that's the right approach," he said.
3 Stefan Kanfer (1993) *The Last Empire: De Beers, Diamonds and the World*. New York: Farrar, Straus, Giroux, 30-1.
4 Ibid., 42.
5 Kimberley Process Statistics Website: https://kimberleyprocessstatistics.org/static/pdfs/public_statistics/2011/2011GlobalSummary.pdf, accessed July 9, 2013.
6 Ibid.

2 SUPPLY AND DEMAND – THE BUSINESS OF DIAMONDS

1 Adam Smith [1776] *An Inquiry Into the Nature and Causes of the Wealth of Nations*, ed. E. Cannan. London: Methuen, 1904, vol. I,

ch. 4: "Of the Origin and Use of Money": http://oll.libertyfund.org/?option=com_staticxt&staticfile=show.php%3Ftitle=237&chapter=212266&layout=html&Itemid=27, accessed Aug. 30, 2013.

2 In his biography of Ernest Oppenheimer, Theodore Gregory lists the global annual production of diamonds between 1911 and 1956. The total is 404 million carats, or an average of 8.78 million per annum. The figures are not strictly comparable with modern production figures because, until the 1930s, there was no significant use for industrial diamonds. Pre- and post-World War II numbers, therefore, do not compare like with like.

3 Kimberley Process Global Database: https://kimberleyprocessstatistics.org/public_statistics, accessed Aug. 30, 2013.

4 A. Thomas (1996) *Rhodes*. New York: St. Martin's Press, 69.

5 Kanfer, *Last Empire*, 59.

6 Ibid. 116.

7 Mark Twain (1897) *Following the Equator*, ch.69: www.writersmugs.com/books/books.php?book=182&name=Mark_Twain&title=Following_the_Equator, accessed Aug. 30, 2013.

8 B. Roberts (1987) *Cecil Rhodes, Flawed Colossus*. London: Hamish Hamilton, 298.

9 E. J. Epstein (1982) *The Rise and Fall of Diamonds*. New York: Simon & Schuster, 81.

10 J. H. Shenefield & I. M. Stelzer (1993) *The Antitrust Laws: A Primer*. Washington: American Enterprise Institute Press, 1; cited in D. L. Spar (undated) *Managing International Trade and Investment: Casebook*. London: Imperial College Press, 213.

11 Kanfer, *Last Empire*, 317.

12 N. Oppenheimer (March 1999) Speech at the Harvard Business School Global Alumni Conference, quoted in Spar, *Managing International Trade*, 220-3.

13 JCK News. "De Beers Antitrust Class Action Now Final": www.jckonline.com/2012/05/23/de-beers-antitrust-class-action-now-final, accessed Aug. 30, 2013.

14 Epstein, *The Rise and Fall of Diamonds*, 178.

15 M. Sevdermish & A. Mashiah (1995) *The Dealer's Book of Gems and Diamonds*, vol. II. Israel: Kal Printing House, 585.

16 Kimberley Process Statistics: https://kimberleyprocessstatistics.org/static/pdfs/public_statistics/2011/2011GlobalSummary.pdf.

17 J. Tagliabue, "An Industry Struggles to Keep Its Luster," *New York Times*, Nov. 5, 2012.
18 Anita Loos (1925) *Gentlemen Prefer Blondes*. New York: Boni & Liveright, 101.
19 Epstein, *The Rise and Fall of Diamonds*, 10.
20 www.streetdirectory.com/travel_guide/35054/jewelry/why_the_three_stone_diamond_ring_is_so_popular_and_desirable.html, accessed Aug. 30, 2013.
21 www.forevermark.com/en/A-Forevermark-Diamond/The-Forevermark-Promise, accessed Aug. 30, 2013.
22 N. Stein, "The De Beers Story: New Cut on an Old Monopoly" *Fortune Magazine*, Feb. 2001
23 www.dailymail.co.uk/news/article-2204566/Russia-diamonds-Source-Siberian-asteroid-crater-supply-world-markets-3-000-years.html, accessed Aug. 30, 2013.

3 BLOOD DIAMONDS

1 I. Smillie, L. Gberie, & R. Hazelton (2000) *The Heart of the Matter: Sierra Leone, Diamonds and Human Security*. Ottawa: Partnership Africa Canada, 5.
2 I. Smillie (April 21, 2007) "Diamonds, the RUF and the Liberian Connection: A Report for the Office of the Prosecutor, Special Court for Sierra Leone," 10; charlestaylortrial.files.wordpress.com/.../p-19-report-ian-smillie.pdf, accessed Aug. 30, 2013.
3 Global Witness (1998) *A Rough Trade*. London: Global Witness, p. 4.
4 Epstein, *The Rise and Fall of Diamonds*, 87.
5 The Democratic Republic of Congo (DRC), briefly known as Zaire, was a Belgian colony. The Republic of Congo, separated from the DRC on its southern and much of its eastern borders by the River Congo, was a French colony.
6 C. Dietrich (2002), *Hard Currency: The Criminalized Diamond Economy of the Democratic Republic of the Congo and its Neighbours*. Ottawa: Partnership Africa Canada, pp. 13 and 17.
7 Ian Smillie (2010), *Blood on the Stone: Greed, Corruption and War in the Global Diamond Trade*. London: Anthem Press, 131.
8 Dietrich, *Hard Currency*, 42.
9 B. Coghlan, P. Ngoy, F. Mulumba, C. Hardy, V. K. Bemo, T.

Stewart, et al. (2007) *Mortality in the Democratic Republic of Congo.* New York: International Rescue Committee.

10 A. Peterman, T. Palermo & C. Bredenkamp (June 2011) "Estimates and Determinants of Sexual Violence Against Women in the Democratic Republic of Congo," *American Journal of Public Health* 101(6): 1060-7.

11 MSNBC, "Liberia's Former President, a Friend to Terror?" *Dateline NBC*, aired July 17, 2005. Ibrahim Bah resurfaced briefly in 2013, living rather openly in Sierra Leone despite a UN travel ban. There, among other things, he was engaged in diamond exports. During an international media uproar, the government of Sierra Leone said that he had been arrested and deported to Senegal. The BBC reported that he never arrived in Senegal and had simply vanished. See "Sierra Leone 'deports' Taylor ally, Ibrahim Bah," BBC, Aug. 7, 2013: www.bbc.co.uk/news/23599725, accessed Sept. 2, 2013.

12 "Al-Qaida Bomber Fazul Abdullah Mohammed Killed," *Guardian*, June 11, 2011: www.guardian.co.uk/world/2011/jun/11/al-qaida-bomber-fazul-abdullah-mohammed-killed.

4 ACTIVISM

1 MONUA Mandate: https://www.un.org/Depts/DPKO/Missions/Monua/monuam.htm, accessed Dec. 17, 2012.

2 Global Witness, *A Rough Trade*, 8.

3 Ibid., 14.

4 R. Fowler & D. Angell (2001) "Angola Sanctions." In R. McRae & D. Hubert (eds.) *Human Security and the New Diplomacy.* Montreal & Kingston: McGill-Queen's University Press, 190.

5 Editorial, *New York Times*, Aug. 8, 1999.

6 M. Rapaport (November 5, 1999) "Blood Money": www.diamonds.net/News/NewsItem.aspx?ArticleID=3347&ArticleTitle=Blood+Money, accessed Dec. 17, 2012.

7 R. Cook, quoted by Tony Hall in Hearing before the Subcommittee on Africa of the Committee on International Relations, House of Representatives, One Hundred Sixth Congress, Second Session, February 15, 2000: https://bulk.resource.org/gpo.gov/hearings/106h/65150.txt, accessed Dec. 17, 2012.

8 UN Security Council, UN S/2000/203, March 10, 2000, Para. 113.
9 Smillie et al., *The Heart of the Matter*, 72.
10 Ibid. 72, citing IRIN Newsbriefs, UN Office for the Coordination of Humanitarian Affairs, Nov. 17, 1999.
11 M. Rapaport (April 7, 2000) "Guilt Trip": www.diamonds.net/News/NewsItem.aspx?ArticleID=3830&ArticleTitle=Guilt+Trip, accessed Dec. 18, 2012.

5 REGULATION

1 WTO: www.wto.org/english/news_e/news03_e/goods_council_26fev03_e.htm, accessed Aug. 30, 2013.
2 The complete definition is as follows: "Conflict diamonds means rough diamonds used by rebel movements or their allies to finance conflict aimed at undermining legitimate governments, as described in relevant United Nations Security Council (UNSC) resolutions insofar as they remain in effect, or in other similar UNSC resolutions which may be adopted in the future, and as understood and recognised in United Nations General Assembly (UNGA) Resolution 55/56, or in other similar UNGA resolutions which may be adopted in future." The full KPCS document is available at www.kimberleyprocess.com/web/kimberley-process/kp-basics.
3 A secretariat of sorts was eventually created in 2012 and is discussed in the following chapter.
4 See, for example, the 2010 UNGA discussion about the Kimberley Process: www.un.org/News/Press/docs/2010/ga11039.doc.htm, accessed Jan. 4, 2013.

6 POWER AND POLITICS

1 Kimberley Process, "Chair's Notice: Review Mission to the Republic of Congo and a Revised List of Participants, 9 July 2004": author's possession.
2 Kimberley Process, "KPCS: Report of the Review Visit to the Democratic Republic of the Congo, Oct. 10-16, 2004": author's possession.

3 Kimberley Process, "Final Report: Review Visit to the Democratic Republic of Congo, 9-14 March, 2009": author's possession.
4 I. Smillie (ed.) (2009) *Diamond Industry Annual Review 2009*. Ottawa: Partnership Africa Canada.
5 S. G. Blore (ed.) (2005) *The Failure of Good Intentions: Fraud, Theft and Murder in the Brazilian Diamond Industry*. Ottawa: Partnership Africa Canada. See also S. G. Blore (ed.) (2006) *Fugitives and Phantoms: The Diamond Exporters of Brazil*. Ottawa: Partnership Africa Canada.
6 Letter to KP Chair President Vyacheslav Shtyrov, June 3, 2005, from Secretary Giles Carriconde Azevedo of Secretaria de Geologia, Mineração e Transformação Mineral of the Government of Brazil: author's possession.
7 "Breakthrough in Conflict Diamond Battle," *Financial Times*, June 14, 2007.
8 Letter from Rodolfo Sanz, Venezuelan Minister of the Popular Power for Mining and Basic Industries, to Bernhard Esau, Kimberley Process Chair, Nov. 3, 2009: author's possession.
9 "Not Just Out of Africa: South America's 'Blood Diamonds' Network," *Time Magazine*, 20 August 20, 2012.
10 Final Communiqué from the Kimberley Process Plenary Meeting, November 30, 2012: www.kimberleyprocess.com/en/2012-final-communique-plenary-washington, accessed Oct. 19, 2013.
11 Author's correspondence with Shawn Blore, April 25, 2013.
12 "Congo Fails to Curb Diamond Smuggling," *Times of India*, August 10, 2011.
13 Arrests were made in May 2013, although at the time of writing it was unclear what percentage of the stolen goods had been recovered. See "Belgian Diamond Theft," *New York Times*, June 16, 2013.
14 J. Miklian, "Rough Cut," *Foreign Policy*, January/February 2013.
15 Chaim Even-Zohar, "The 'Secret Diamond Mines' of Lebanon," *Diamond Intelligence Briefs*, May 20, 2009.
16 P. Greenhalgh (1985) *West African Diamonds 1919-83: An Economic History*. Manchester: Manchester University Press, 157.
17 United Nations, "Midterm Report of the Panel of Experts on Liberia" (S/2011/367), June 22, 2011, 16.
18 Trade data from KP Statistics website, accessed Jan. 23, 2013.
19 Chaim Even-Zohar, "India's Diamond Woes," *Diamond Intelligence Briefs*, May 16, 2013.

20 Kimberley Process, "Report of the Review Visit of the Kimberley Process Certification Scheme to the United Arab Emirates, 14 to 18 January 2008": author's possession.
21 C. Even-Zohar, "India: Going Straight on Round-Tripping," *Diamond Intelligence Briefs*, February 6, 2013.
22 The only other country with dedicated UNSC diamond sanctions after 2005 was Côte d'Ivoire.
23 Kimberley Process, "Report of the Review Visit of the Kimberley Process to the Republic Of Angola, 11-15, October 2005": author's possession.
24 Statement issued by the Council of European Union Foreign Ministers, Jan. 26, 2009.
25 www.scoop.co.nz/stories/WL0908/S00699.htm, accessed Aug. 30, 2013.
26 David Smith, "Zimbabwean Government Bank Balance Down to $217," *Guardian*, Jan. 30, 2013.
27 Parliament of Zimbabwe, First Report of the Portfolio Committee on Mines and Energy on Diamond Mining (with special reference to Marange Diamond Fields), 2009–2013; presented to Parliament, June 2013 (S.C.4, 2012).
28 Nehanda Radio website: http://nehandaradio.com/2013/06/25/diamond-mining-in-zimbabwe-the-chininga-report, accessed June 27, 2013.
29 Kimberley Process, "Final Communiqué from the Kimberley Process Plenary Meeting," November 30, 2012, Washington, D.C.
30 My emphasis.
31 Global Witness. "Why We Are Leaving the Kimberley Process – A Message from Global Witness Founding Director Charmian Gooch," December 5, 2011: www.globalwitness.org/library/why-we-are-leaving-kimberley-process-message-global-witness-founding-director-charmian-gooch, accessed Jan. 23, 2013.
32 M. Rapaport (December 2012) "Moral Clarity and the Diamond Industry," *Rapaport Magazine*, available online: www.diamonds.net/Magazine/Article.aspx?ArticleID=41814&RDRIssueID=101&ArticleTitle=Moral+Clarity+and+the+Diamond+Industry, accessed Aug. 30, 2013.

7 DEVELOPMENT

1 B. Keller, "Mandela's Mogul," *New York Times Magazine*, Jan. 27, 2013.
2 Kimberley Process statistics for 2012, Kimberley Process Statistics website: https://kimberleyprocessstatistics.org/public_statistics.
3 Transparency International: http://cpi.transparency.org/cpi2012/results/, accessed Aug. 30, 2013.
4 UNDP Human Development Index 2011: http://hdr.undp.org/en/statistics/hdi, accessed Aug. 30, 2013.
5 Partnership Africa Canada, "An Interview with the Mineral Resources Minister," *Sierra Leone Diamond Industry Annual Review 2006*, Ottawa: PAC, 2006.
6 Global Witness Publishing Inc. and Partnership Africa Canada, *Rich Man, Poor Man: Development Diamonds and Poverty Diamonds – The Potential for Change in the Artisanal Alluvial Diamond Fields of Africa*. Ottawa and Washington: Global Witness Publishing Inc. and Partnership Africa Canada, 2004.
7 Despite past failures, new investors seem keen to break the run of insolvencies. Deepak International Ltd. signed an agreement with the Northwest Territories government in 2013 to set up a new cutting and polishing operation in Yellowknife, buying facilities vacated by previous diamond investors.
8 Nicky Oppenheimer, interviewed in *Mazal U'Bracha Diamonds*, August 2000. Ironically, De Beers later made an arrangement with a Canadian cutting and polishing firm in Sudbury, Ontario, to take up to 10 percent of the output of its Victor Mine near Attawapiskat, Ontario. In this case De Beers agreed to let the company have stones of more than 1 carat. Of the company's first 29 employees, however, 23 were from Vietnam, not Ontario.
9 "State-backed Diamond Beneficiation Firm to be Liquidated," *Mining Weekly*, Jan. 18, 2013.
10 Gavin Lewis, "Gauteng's R144m Loan Loss to ANC MP Firm," Blogpost, August 30 2012: http://da-gpl.co.za/?p=1224, accessed Jan. 28, 2013.

8 LOOSE ENDS

1 Several countries have adopted a lower threshold of $10,000 or its equivalent.
2 R. Bates, "Undisclosed Synthetic Diamonds Appearing on Market," *JCK Online*, May 21, 2012: www.jckonline.com/2012/05/21/undisclosed-synthetic-diamonds-appearing-on-market, accessed Feb. 2, 2013.
3 www.diamonds.net/News/NewsItem.aspx?ArticleID=41967&ArticleTitle=A+Look+Ahead+%E2%80%8E, accessed June 27, 2013.
4 "JVC Honors Runci": http://news.centurionjewelry.com/articles/view/industry-news-honors-at-jvc-jsa-events-contest-deadline-nears-jtv-founder-d, accessed Feb. 2, 2012.

Selected readings

Chapter 1 There are many technical books and articles on the geological origin of diamonds. The most readable treatment can be found in Kevin Krajik's *Barren Lands: An Epic Search for Diamonds in the North American Arctic* (New York: Henry Holt, 2001). Peter Greenhalgh describes the evolution of West African diamonds in *West African Diamonds 1919-1983: An Economic History* (Manchester: Manchester University Press, 1985). The South African diamond story has been told many times; one of the best versions is Stefan Kanfer's *The Last Empire: De Beers, Diamonds and the World* (New York: Farrar Straus Giroux, 1993).

Chapter 2 Anthony Thomas' 1996 *Rhodes* (New York: St. Martin's Press) contains considerable detail on the Cecil Rhodes diamond story. The Oppenheimer takeover saga is detailed in Theodore Gregory, *Ernest Oppenheimer and the Economic Development of Southern Africa* (Cape Town: Oxford University Press, 1962), in Epstein's *The Rise and Fall of Diamonds* and Kanfer's *The Last Empire.*

Chapter 3 The IDSO story and others about diamond smuggling during the 1950s can be found in Ian Fleming's semi-factual *The Diamond Smugglers* (London: Jonathan Cape, 1957) and the more trustworthy *Sir Percy Sillitoe* by A. W. Cockerill (London: W. H. Allen, 1975). Sillitoe wrote his own more guarded memoir, *Cloak Without Dagger* (London: Cassel, 1955), and Fred Kamil's dubious memoir, *The Diamond Underworld*, gives a more grassroots perspective (London:

Allen Lane, 1975). See also *Diamonds are Dangerous,* by J. H. du Plessis (New York: John Day & Co., 1960). Sierra Leone's collapse is described in William Reno's *Corruption and State Politics in Sierra Leone* (Cambridge: Cambridge University Press, 1995) and Lansana Gberie's *A Dirty War in West Africa* (Bloomington and Indianapolis: Indiana University Press, 2005). Details on the criminalization of the diamond industry and the horrors of the war can be found in Gberie and in my own book, *Blood on the Stone: Greed, Corruption and War in the Global Diamond Trade* (London: Anthem Press, 2010). See also *Blood Diamonds* by Greg Campbell (London: Westview Press, 2002; updated 2012), which focusses primarily on Sierra Leone. Charles Taylor's ascendency is described in Mark Huband's *The Liberian Civil War* (London: Frank Cass, 1998) and Stephen Ellis' *The Mask of Anarchy* (London: C. Hurst & Co., 1999). Tony Hodges' 2001 book, *Angola from Afro-Stalinism to Petro-Diamond Capitalism* (London: James Currey) provides a good background to the Angolan conflict and the role of natural resources. There are many excellent books about the DRC. *King Leopold's Ghost* by Adam Hochschild (Boston: Houghton Mifflin, 1998) chronicles the colonial era, and Michaela Wrong's 2000 book, *In the Footsteps of Mr. Kurtz* (London: Fourth Estate), discusses the Mobutu era and its aftermath. Partnership Africa Canada has many diamond-related studies on Sierra Leone, Liberia, Angola, and the DRC on its website: www.pacweb.org (accessed Aug. 30, 2013; this date accessed applies to all other URLs in this section unless otherwise noted). Douglas Farah tells the al Qaeda story in *Blood from Stones* (New York: Broadway Books, 2004). In *Wars of Plunder* (New York: Columbia University Press, 2012), Philippe Le Billon unpacks various debates about whether diamonds are a general or situational resource curse.

Chapter 4 The Global Witness and Partnership Africa Canada reports described in this chapter, along with many more, are available on their websites: www.globalwitness.org and www.pacweb.org respectively. The Fowler Report to the United Nations on Angola can be found online by searching for UN S/2000/203, and the Expert Panel Report on Sierra Leone can be found by searching for UN S/2000/1195. Other information and reports of subsidiary bodies of the UN Security Council can be found at www.un.org/en/sc/subsidiary. I have written about the NGO "blood diamond" campaign in my own book, *Blood on the Stone*, and in *Global Accountabilities*, ed. A. Ebrahim & E. Weisband (Cambridge: Cambridge University Press, 2007) and *Business Regulation and Non-State Actors*, ed. D. Reed, P. Utting & A. Mukherjee-Reed (London: Routledge, 2012).

Chapter 5 This chapter describes the Kimberley Process Certification Scheme, details of which can be found on the Kimberley Process website at www.kimberleyprocess.com/web/kimberley-process/home. This site also describes the KP working groups and many of the administrative decisions that have been taken since the KPCS began operation. The Kimberley Process statistical website can be found at https://kimberleyprocessstatistics.org/public_statistics.

Chapter 6 The website of Partnership Africa Canada (www.pacweb.org) is a treasure house of detailed information about diamonds in several countries. Between 2004 and 2007, PAC produced semi-annual reports on diamonds in the Democratic Republic of Congo and Angola, and there are detailed chapters on both countries in its *Diamond Industry Annual Reviews* of 2008 and 2009. Full details on the Brazil case can be found in two additional PAC reports available on the website: *The Failure of Good Intentions: Fraud, Theft and Murder in the Brazilian Diamond Industry* (2005) and *Fugitives and Phantoms: The Diamond Exporters of Brazil* (2006). The Venezuelan saga is detailed in PAC's *The Lost World: Diamond Mining and Smuggling in Venezuela* (2006) and in various issues of *Other Facets*, all found on its website. PAC also has several reports on Zimbabwe. A good 2012 Human Rights Watch report on Angola can be found at www.hrw.org/reports/2012/05/20/if-you-come-back-we-will-kill-you. Although the open KP Statistics Website does not make all trade data available, it provides interesting information: https://kimberleyprocessstatistics.org/public_statistics.

Chapter 7 The Endiama website provides details on Angolan investment requirements: www.endiama.co.ao/english/index.php. Much has been written about impact benefit agreements. The IBA Research Network has a good bibliography at www.impactandbenefit.com/Research. *Rich Man, Poor Man* can be found at www.pacweb.org/en/diamonds-research-pubs. *Guyana's Diamond Tracking System: A Model for Artisanally Mined Diamonds* and other aspects of the work of the Diamond Development Initiative can be found on the DDI website at www.ddiglobal.org.

Chapter 8 Information on the Responsible Jewellery Council can be found on its website at www.responsiblejewellery.com. Information on the Financial Action Task Force can be found on its website at www.fatf-gafi.org/topics/fatfrecommendations/documents/the40recommendationspublishedoctober2004.html.

Index